Condition Monitoring Using Computational Intelligence Methods

Tshilidzi Marwala

Condition Monitoring Using Computational Intelligence Methods

Applications in Mechanical and Electrical Systems

Springer

Tshilidzi Marwala
Faculty of Engineering and the Built Environment
University of Johannesburg
Auckland Park
South Africa

ISBN 978-1-4471-2379-8 e-ISBN 978-1-4471-2380-4
DOI 10.1007/978-1-4471-2380-4
Springer London Dordrecht Heidelberg New York

British Library Cataloguing in Publication Data
A catalogue record for this book is available from the British Library

Library of Congress Control Number: 2012931151

Printed on acid-free paper

Springer is part of Springer Science+Business Media (www.springer.com)

Foreword

Condition monitoring is a process of monitoring a system by studying certain selected parameters in such a way that significant changes of those parameters are related to a developing failure. It is a significant type of predictive maintenance and permits maintenance to be intelligently managed. The supply of reliable electricity and machines that are safe to operate are cornerstones of building a caring society. In this regard, the book *Condition Monitoring Using Computational Intelligence Methods* Tshilidzi studies mechanical and electrical systems. It is usually desirable that the condition monitoring process be fully automated. Furthermore, it is desirable for this automation process to be intelligent. Many techniques have been developed to capacitate processes with intelligent decision-making and chief amongst this is artificial intelligence which is also known as computational intelligence. This paradigm has made it possible to design robots that are not only able to perform routine tasks but are able to perform tasks that are unexpected. This has capacitated robots to operate under highly uncertain environments.

The book *Condition Monitoring Using Computational Intelligence Methods* introduces techniques of artificial intelligence for condition monitoring of mechanical and electrical machines. It also introduces the concept of on-line condition monitoring as well as condition monitoring in the presence of sensor failures. Furthermore, it introduces various signals that can be used for condition monitoring.

This book is useful for graduate students, researchers and practitioners.

Harare — Arthur G.O. Mutambara, D.Phil.

Preface

Condition Monitoring Using Computational Intelligence Methods introduces the concept of computational intelligence to monitoring the condition of mechanical and electrical systems. The book implements multi-layer perception neural networks, Bayesian networks, a committee of neural networks, Gaussian mixture models, hidden Markov models, fuzzy systems, ant colony optimized rough sets models, support vector machines, the principal component analysis and extension networks to create models that estimate the condition of mechanical and electrical systems, given measured data. In addition, the Learn ++ method is applied to create an on-line computational intelligence device which can adapt to newly acquired data. Furthermore, auto-associative neural networks and the genetic algorithm are used to perform condition monitoring in the presence of missing data. The techniques that used on datasets were pseudo-modal energies, modal properties, fractal dimensions, mel-frequency cepstral coefficients, and wavelet data.

This book makes an interesting read and opens new avenues and understanding in the use of computational intelligence methods for condition monitoring.

University of Johannesburg

Tshilidzi Marwala, Ph.D.

Acknowledgements

I would like to thank the following former and current graduate students for their assistance in developing this manuscript: Lindokuhle Mpanza, Ishmael Sibusiso Msiza, Nadim Mohamed, Dr. Brain Leke, Dr. Sizwe Dhlamini, Thando Tettey, Bodie Crossingham, Professor Fulufhelo Nelwamondo, Vukosi Marivate, Dr. Shakir Mohamed, Dalton Lunga and Busisiwe Vilakazi. I also thank Mpho Muloiwa for assisting me in formatting the book. Thanks also go to the colleagues and practitioners that have collaborated directly and indirectly to writing of the manuscript. In particular, we thank Dr. Ian Kennedy, Professor Snehashish Chakraverty and the anonymous reviewers for their comments and careful reading of the book. Finally I wish to thank my supervisors Dr. Hugh Hunt, Professor Stephan Heyns and Professor Philippe de Wilde.

I dedicate this book to Jabulile Manana as well as my sons Thendo and Khathutshelo.

University of Johannesburg — Tshilidzi Marwala, Ph.D.

Contents

Chapter 1
Introduction to Condition Monitoring in Mechanical and Electrical Systems

1.1 Introduction

A procedure for monitoring and identifying faults in systems is of vital importance in electrical and mechanical engineering. For instance, aircraft operators must be sure that their aircrafts are free from cracks. Cracks in turbine blades lead to catastrophic failure of aircraft engines and must be detected as early as possible. Bridges nearing the end of their useful life must be assessed for their load-bearing capacities.

Many techniques have been employed in the past for monitoring the condition of systems. Some techniques are visual (e.g., dye penetrating methods) and others use sensors to detect local faults (through acoustics, magnetics, eddy currents, thermal fields, and radiographs). These methods are time consuming and cannot show that a structure is fault-free without testing the entire structure in minute detail. Furthermore, if a fault is buried deep within the structure, it may not be visible or detectable using these localised techniques. The need to detect faults in complicated structures has led to the development of global methods, which can use changes in the measured data of the structure as a basis for fault detection (Doebling et al. 1996; Marwala 1997, 2001).

Yadav et al. (2011) implemented an audio signature for monitoring of the condition of internal-combustion engines using a Fourier transform and a correlation approach to categorize whether the engine was healthy or faulty.

He et al. (2009) applied Principal Component Analysis (PCA) for monitoring the condition of an internal-combustion engine through sound and vibration analysis of an automobile gearbox. They found that their technique was effective for the monitoring of machine conditions. Bouhouche et al. (2011) also successfully applied Principal Component Analysis (PCA) and a self-organization map (SOM) for monitoring the condition of a pickling process. A comparison of self-organization maps, the traditional PCA, and the mixture of PCA-SOM was made using the data obtained from a real pickling process. The hybrid method was better than the PCA but not better than the SOM. Loutas et al. (2011) applied vibration,

T. Marwala, *Condition Monitoring Using Computational Intelligence Methods: Applications in Mechanical and Electrical Systems*, DOI 10.1007/978-1-4471-2380-4_1,

acoustic emission, and oil debris analysis for the on-line condition monitoring of rotating machinery. Multi-hour tests were performed on healthy gearboxes until they were damaged and on-line monitoring methods were examined. A number of parameters/features were extracted from the time and frequency domain as well as from wavelet-based signal processing. A PCA was used to condense the dimensionality of the data, while an independent component analysis was used to categorize the independent components in the data and correlate them with the different fault modes of the gearbox. The combination of vibration, acoustic emission, and oil debris data increased the diagnostic capability and dependability of the condition monitoring technique.

Park et al. (2011) successfully implemented electro-mechanical impedance-based wireless for the condition monitoring of the de-bonding of Carbon Fiber Reinforced Polymer (CFRP) from laminated concrete structures. The CFRP-reinforced concrete samples were made and impedance signals were measured from the wireless impedance sensor node with different de-bonding conditions between the concrete and the CFRP. Cross correlation data analysis was used to estimate the changes in impedance measured at the patches due to the de-bonding conditions. The results indicated that impedance-based wireless Structural Health Monitoring (SHM) can be implemented successfully for monitoring the de-bonding of CFRP laminated concrete structures.

Murthy et al. (2011) applied condition monitoring analysis on surveillance videos of insulators of electric power lines. This was conducted by monitoring both the voltage and leakage flow. The method applied a Wavelet Coefficient Differentiator (WCD) to analyze the signal. This method was found to give good results in less time, when compared to traditional approaches.

Tian et al. (2011) applied a condition monitoring technique in wind turbine components to decrease the operation and maintenance costs of wind power generation systems. Their maintenance technique estimated the failure probability values and a simulated study demonstrated the advantage of the proposed technique for reducing the maintenance cost.

Al-Habaibeh et al. (2002) applied Taguchi's method to provide an extensive experimental and analytical evaluation of a previously presented approach for the systematic design of condition monitoring systems for machining operations. When the technique was evaluated on tool faults in end-milling operations, it showed that it can successfully detect faults. Zhu et al. (2009) applied wavelet analysis (which is a non-stationary signal processing technique) for the condition monitoring of tools. This study successfully reviewed five processes; namely; time-frequency analysis of the machining signal, signal de-noising, feature extraction, singularity analysis for tool state estimation and density estimation for the classification of tool wear.

Vilakazi et al. (2005) applied condition monitoring to bushings and used Multi-Layer Perceptrons (MLPs), Radial Basis Functions (RBFs) and Support Vector Machine (SVM) classifiers. The first level of their framework determined if the bushing was faulty or not, while the second level determined the type of fault. The diagnostic gases in the bushings were analyzed using dissolved gas analysis. The MLP gave better accuracy and training time than SVM and RBF did.

In addition, an on-line bushing condition monitoring approach, which could adapt to newly acquired data, was introduced. This approach could accommodate new classes that were introduced by incoming data. The approach was implemented with an incremental learning algorithm that used MLP. The testing results improved from 67.5% to 95.8% as new data were introduced and the testing results improved from 60% to 95.3% as new conditions were introduced. On average, the confidence value of the framework about its decision was 0.92.

Bartelmus and Zimroz (2009) applied features for monitoring the condition of gearboxes in non-stationary operating conditions. The method used a simple regression equation to estimate diagnostic features. The technique was found to be very fast, simple, dynamic, and intuitive.

Vilakazi and Marwala (2007) successfully applied an incremental learning method to the problem of the condition monitoring of electrical systems. Two incremental learning methods were applied to the problem of condition monitoring. The first technique used the incremental learning ability of Fuzzy ARTMAP (FAM) and explored whether ensemble methods could improve the performance of the FAM. The second technique used Learn++ that applied an ensemble of MLP classifiers. Later, Vilakazi and Marwala (2009) applied a novel technique to the condition monitoring of bushing faults by using FAM. FAM was introduced for bushing condition monitoring because it has the capability to incrementally learn information as the information is made available. An ensemble of classifiers was used to improve the classification accuracy of the systems. The results demonstrated that a FAM ensemble gave an accuracy of 98.5%. Additionally, the results showed that a FAM could update its knowledge in an incremental fashion without forgetting previously learned information.

Nelwamondo and Marwala (2007) successfully applied several methods to handle missing data, which included a novel algorithm that classifies and regresses in a condition monitoring problem having missing data.

Miya et al. (2008) applied an Extension Neural Network (ENN), a Gaussian Mixture Model (GMM) and a Hidden Markov Model (HMM) for condition monitoring of bushings. The monitoring process had two-stages: (1) detection of whether the bushing was faulty or normal and (2) a classification of the fault. Experiments were conducted using data from a Dissolved Gas-in-oil Analysis (DGA) collected from bushings and based on the IEEEc57.104; IEC60599 and IEEE production rates methods for Oil-Impregnated Paper (OIP) bushings. It was observed from experimentation that there was no difference in major classification between ENN and GMM in the detection stage with classification rates of 87.93% and 87.94% respectively, outperforming HMM which achieved only 85.6%. Moreover, the HMM fault diagnosis surpassed those of ENN and GMM with a classification success of 100%. For the diagnosis stage, the HMM was observed to outperform both the ENN and the GMM with a 100% classification success rate. ENN and GMM were considerably faster at training and classification, whereas HMM's training was time-consuming for both the detection and diagnosis stages.

Booth and McDonald (1998) used artificial neural networks for the condition monitoring of electrical power transformers while Pedregal and Carnero (2006) applied state space models that used a Kalman filter and vibration data for the condition monitoring of turbines.

From the literature review above, there are few key terminologies that are emerging. These are data measurement, signal processing, and machine learning (e.g. SVM and neural networks). From these key variables this book constructs a generalized condition monitoring framework, which is the subject of the following section.

1.2 Generalized Theory of Condition Monitoring

The generalized theory of condition monitoring is illustrated in Fig. 1.1, with one device per box. This figure shows in the first box that there is a data acquisition device, whose primary function is to acquire data from the system. Examples of these would include measurement devices such as thermometers, accelerometers, or strain gauges.

The second box in the figure contains the data analysis device, whose function is to analyze the acquired data. Many methods, some of which will be described in Chap. 2, have been proposed in this regard. The methods include using wavelets, the Fourier transform, and the Wagner-Ville distribution.

In the fourth box, feature selection is a process where specific aspects of the data, which are good indicators of faults in the structure, are identified and quantified.

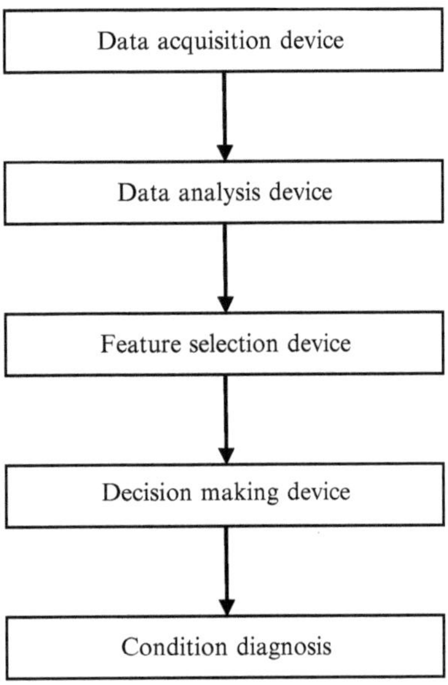

Fig. 1.1 Condition monitoring framework

Methods that have been developed include independent component analysis and the principal component analysis, which will also be described in Chap. 2.

The decision making device is an infrastructure whose primary function is to take the features and interpret these features. Methods that have been used include the Multi-Layer Perceptrons and Radial Basis functions, which will be described in Chap. 3, a committee of networks which will be described in Chap. 4, a Bayesian network which will be described Chap. 5, a support vector machine which will be described in Chap. 6, a fuzzy system which will be described in Chap. 7, and rough sets system which will be described in Chap. 8. The outcome of the decision making device is the identification of faults.

In implementing the procedures in Fig. 1.1, Gunal et al. (2009) used the motor current as the data, notch-filtering in the analysis, with feature devices, and finally, used popular classifiers as the decision making device to establish whether the induction motor was healthy or not.

Loutas et al. (2011) applied vibration, oil debris and acoustic emission as the data acquisition and analysis device, principal component analysis as the feature extraction device and heuristics rules as the decision making device to monitor the condition of a rotating machine.

Elangovan et al. (2010) used a continuous acquisition of signals from sensor systems, extracted features using statistical and histogram methods and used a Bayes classifier for the condition monitoring of single point carbide tipped tool. Zhou et al. (2011a) used position sensors for the data acquisition device and an ensemble empirical mode decomposition method for gearbox condition monitoring. Garcia-Escudero et al. (2011) used motor line current as data and a Fast Fourier Transform as the feature selection device with robust quality control based on multivariate control charts as a decision making device, making early detection of broken rotor bars in induction motors possible.

1.3 Stages of Condition Monitoring

The aim of the condition monitoring process is to estimate the state of health of a structure or machine from measured data. The state of health of the structure or machine can be estimated through the five stages which are shown in Fig. 1.2. The first stage in fault estimation is the *detection* of the presence or the absence of a fault. Zhou et al. (2011b) used feature identification for industrial fault detection, while Zhang and Huang (2011) successfully detected faults in hybrid fuel cells. Zhu et al. (2011) detected faults for a class of nonlinear systems, while Hussain and Gabbar (2011) detected faults for real time gears based on a pulse shape analysis.

The next stage of fault estimation is *fault classification* which, in this chapter, is defined as more than just classifying the presence or the absence of fault but includes the nature of the fault (e.g., extent and type).

Kim et al. (2010a, b) used Support Vector Machines for classifying fault types in rotating machines, while Lin et al. (2010) applied a hybrid of rough sets and neural networks for classifying the types of faults in transmission lines. Thai and

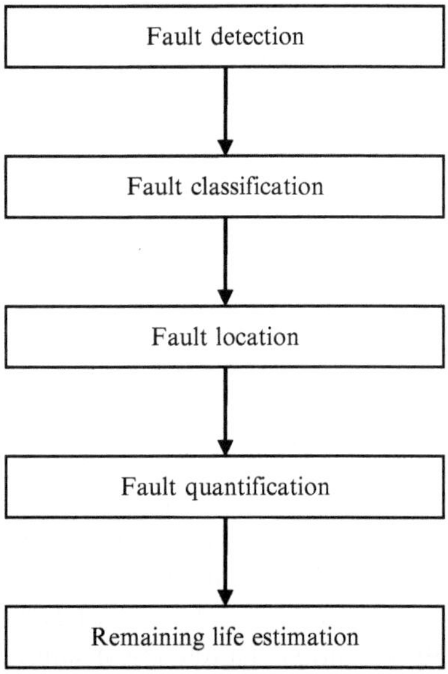

Fig. 1.2 Fault stages

Yuan (2011) applied neuro-fuzzy techniques for classifying the types of transmission line faults. Abdel-Latief et al. (2003) applied statistical functions and neural networks for the classification of fault types in power distribution feeders.

The next stage in fault estimation is the identification of the *location* of the fault. Jayabharata Reddy and Mohanta (2011) applied a modular method for the location of arcing and non-arcing faults on transmission lines. Jain et al. (2009) used terminal data for fault location in double circuit transmission lines while Xie et al. (2009) applied ant colony optimization for the location of faults and Khorashadi-Zadeh and Li (2008) applied neural networks to the location of faults on medium voltage cables.

The next stage in fault estimation is the *quantification* of the magnitude of the fault. Treetrong (2011a) applied a higher-order spectrum technique to quantify the degree of the fault in industrial electric motors. Riml et al. (2010) quantified the faults arising from disregarding the standardised procedures for photographing faces, and Sinha (2009) studied trends in fault quantification of rotating machines.

The last stage in fault estimation is to finally *estimate the remaining life* of the structure that is being monitored. Zio and Peloni (2011) applied a particle filtering method to estimate the remaining useful life of nonlinear components, while Butler and Ringwood (2010) also applied a particle filtering technique for estimating the remaining useful life of abatement equipment which is used in semiconductor manufacturing. Yanagawa et al. (2010) estimated the remaining life of the hydro-turbine in a hydro-electric power station and Kim et al. (2010a, b) applied computer simulation for estimating the remaining life of a level luffing crane component. Gedafa et al. (2010) used surface deflection to estimate the remaining service life of flexible pavements while Garvey et al. (2009) applied pattern recognition methods to estimate the remaining useful life of bottomhole assembly tools, and Pandurangaiah et al. (2008) developed a technique for estimating the remaining life of power transformers.

1.4 Data Used for Condition Monitoring

There are four main domains in which data may be represented: time domain, modal domain, frequency domain, or time-frequency domain (Marwala 2001). Raw data are measured in the time domain. From the time domain, Fourier transforms can be used to transform the data into the frequency domain. From the frequency domain data, and sometimes directly from the time domain, the modal properties may be extracted. All of these domains are reviewed in this chapter. Theoretically, they contain similar information, but in reality this is not necessarily the case.

1.4.1 Time Domain Data

Time domain data is unprocessed data measured over historical time. Normally when such data are used, some form of statistical analysis such as variance and means are used. Tao et al. (2007) applied a time-domain index for the condition monitoring of rolling element bearings. They presented a new statistical moment, derived from the Rényi entropy and compared it to other statistical parameters such as kurtosis and stochastic resonance.

Andrade et al. (2001) applied a new method to the time-domain vibration condition monitoring of spur gear that used a Kolmogorov-Smirnov test. This technique was performed by using a null hypothesis that assumed that the Cumulative Density Function (CDF) of the target distribution is statistically similar to that of a reference distribution. This demonstrated that, in spite of its simplicity, the Kolmogorov-Smirnov test is a powerful technique that successfully classifies different vibration signatures, permitting its safe use as a condition monitoring technique.

Zhang and Suonan (2010) applied the time domain method for fault location in Ultra High Voltage (UHV) transmission lines, while Haroon and Adams (2007) applied the time and frequency domain nonlinear system characterization for the mechanical fault identification in the suspension systems of ground vehicles.

1.4.2 Modal Domain Data

The modal domain data are articulated as natural frequencies, damping ratios and mode shapes. These will be described in detail in Chap. 2. The most widespread method of extracting the modal properties is by using modal analysis (Ewins 1995). This technique has been applied for fault identification. So, in this chapter, both techniques are reviewed.

1.4.2.1 Natural Frequencies

The analysis of shifts in natural frequencies caused by the change in condition of structures or machines has been used to identify structural faults. Because the changes in natural frequencies caused by average fault levels are of small magnitudes, an accurate method of measurement is vital for this technique to be successful. This issue limits the level of fault that natural frequencies can identify to that of high magnitudes (Marwala 2001).

Cawley and Adams (1979) used changes in natural frequencies to detect the health condition of composite materials. To calculate the ratio between frequency shifts for two modes, they implemented a grid between possible fault points and assembled an error term that related measured frequency shifts to those predicted by a model based on a local stiffness reduction. Farrar et al. (1994) applied the shifts in natural frequencies to monitor the condition on an I-40 bridge and observed that the shifts in the natural frequencies were not adequate to be used for detecting faults of small magnitudes. To improve the accuracy of the natural frequency technique, it was realized that it was more feasible to conduct the experiment in controlled environments where the uncertainties in measurements were relatively low. In one such experiment, a controlled environment used resonance ultrasound spectroscopy to measure the natural frequencies and determine the out-of-roundness of ball bearings (Migliori et al. 1993).

Faults in different regions of a structure may result in different combinations of changes in the natural frequencies. As a result, multiple shifts in the natural frequencies can indicate the location of fault. Messina et al. (1996) successfully used the natural frequencies to locate single and multiple faults in a simulated 31 bar truss and tabular steel offshore platform. A fault was introduced into the two structures by reducing the stiffness of the individual bars by up to 30%. This method was experimentally validated on an aluminum rod test structure, where the fault was introduced by reducing the cross-sectional area of one of the members from 7.9 to 5.0 mm.

He et al. (2010) applied natural traveling-wave frequencies to locate faults in electrical systems. They achieved this by analyzing the transient response of a capacitor voltage transformer and its effect on the spectra of fault traveling waves. Xia et al. (2010) used changes in natural frequencies to locate faults in mixed overhead-cable lines. They achieved this by implementing a method based on natural frequencies and an Empirical Mode Decomposition (EMD) for mixed overhead-cable line fault identification. They used EMD to decompose a signal to first identify the necessary part of the travelling wave before extracting the principal component of natural frequency spectra of the traveling wave. The natural frequency's spectra were then analyzed to remove the principal component of natural frequencies spectra of the faulty traveling wave and thereby identify the fault locations. A simulated study showed that their technique can reasonably solve the spectra aliasing problem in a fault location exercise.

Other successful applications of natural frequencies for condition monitoring include that of Huang et al. (2009) in ultra-high voltage transmission lines and Luo et al. (2000) in the real-time condition monitoring in machining processes.

To improve the use of the natural frequencies to detect faults of small magnitude, high-frequency modes, which are associated with local responses, may be used. There are two main problems with working with high frequency modes. First of all, modal overlap is high; and secondly, high frequency modes are more sensitive to environmental conditions than the low frequency modes are.

1.4.2.2 Damping Ratios

The use of damping ratios to detect the presence of fault in structures has been applied mostly to composite materials. Lifshitz and Rotem (1969) studied the changes caused by faults to dynamic moduli and the damping of quartz particle filled resin specimens having either epoxy or polyester as the binder. They introduced a fault by applying a static load and observed that damping was more sensitive to the fault than to the dynamic moduli. Schultz and Warwick (1971) also observed that damping was more sensitive to faults than the use of natural frequencies in glass-fiber-reinforced epoxy beams. Lee et al. (1987) studied the damping loss factors for various types of fault cases in Kevlar/epoxy composite cantilevered beams. They found that damping changes were difficult to detect when a fault was introduced by milling two notches of less than 5% of the cross-sectional area. However, they also found that the damping factors were sensitive when a fault was introduced through the creation of delamination by gluing together two pieces of glass/epoxy and leaving particular regions unglued.

Lai and Young (1995) observed that the delamination of graphite/epoxy composite materials increased the damping ratio of the specimen. They also observed that the damping ratios decrease significantly when the specimen is exposed to humid environments for a prolonged period.

1.4.2.3 Mode Shapes

Mode shapes are the properties of the structure that show the physical topology of a structure at various natural frequencies. They are, however, computationally expensive to identify; are susceptible to noise due to modal analysis; do not take into account the out-of-frequency-bandwidth modes; and they are only applicable to lightly damped and linear structures (Marwala 2001; Doebling et al. 1996). However, the mode shapes are easy to implement for fault identification; are most suitable for detecting large faults; are directly linked to the shape of the structure; and focus on vital properties of the dynamics of the structure (Marwala 2001; Doebling et al. 1996).

West (1984) applied the Modal Assurance Criterion (MAC) (Allemang and Brown 1982), a technique that is used to measure the degree of correlation between two mode shapes, to locate faults on a Space Shuttle Orbiter body flap. A fault was introduced using acoustic loading. The mode shapes were partitioned and changes in the mode shapes across various partitions were compared (Marwala 2001).

Kim et al. (1992) applied the Partial MAC (PMAC) and the Co-ordinate Modal Assurance Criterion (COMAC) presented by Lieven and Ewins (1988) to identify the damaged area of a structure. Mayes (1992) used the mode shape changes for fault localization by using a Structural Translational and Rotational Error Checking which was calculated by taking the ratios of the relative modal displacements from faulty and healthy structures as a measure of the accuracy of the structural stiffness between two different structural degrees of freedom (Marwala 2001).

Salawu (1995) introduced a global damage integrity index, based on a weighted-ratio of the natural frequencies of faulty to healthy structures. The weights were used to indicate the sensitivity of each mode to fault.

Kazemi et al. (2010) successfully applied the modal flexibility variation for fault identification in thin plates. They conducted this experiment by using the variation of modal flexibility and the load-deflection differential equation of plate combined with the invariant expression for the sum of transverse load to develop the fault indicator and a neural network to estimate the fault severity of identified parts.

Furthermore, Kazemi et al. (2011) applied a modal flexibility variation method and genetic algorithm trained neural networks for fault identification. They showed the feasibility of the Modal Flexibility Variation method using numerical simulation and experimental tests carried out on a steel plate. Their results indicated that the performance of the procedure was good.

Liguo et al. (2009) applied modal analysis for fault diagnosis of machines. To assess the legitimacy of modal analysis approaches for fault diagnosis of machines, a simulation study on gearbox was successfully conducted. Ma et al. (2007a, b) successfully used a modal analysis and finite element model analysis of vibration data for fault diagnosis of an AC motor. In particular, they applied a modal analysis technique for the fault identification in an induction motor.

Khosravi and Llobet (2007) presented a hybrid technique for fault detection and modeling based on modal intervals and neuro-fuzzy systems whereas Zi et al. (2005) applied modal parameters for a wear fault diagnosis using a Laplace wavelet.

1.4.3 Frequency Domain Data

The measured excitation and response of a structure can be transformed into the frequency domain using Fourier transforms (Ewins 1995; Marwala 2001). The ratio of the response to excitation in the frequency domain at each frequency is called the frequency response function.

Frequency domain methods are difficult to use in that they contain more information than is necessary for fault detection (Marwala 2001; Ewins 1995). There is also no method to select the frequency bandwidth of interest, and they are usually noisy in the anti-resonance regions. Nevertheless, frequency domain methods have the following advantages (Marwala 2001; Ewins 1995): the measured data comprise the effects of out-of-frequency-bandwidth modes; one measurement offers ample data; modal analysis is not necessary and consequently modal identification errors are circumvented; frequency domain data are appropriate to structures with high damping and modal density.

Sestieri and D'Ambrogio (1989) used frequency response functions to identify faults while D'Ambrogio and Zobel (1994) applied frequency response functions to identify the presence of faults in a truss-structure.

Imregun et al. (1995) observed that the direct use of frequency response functions to identify faults in simulated structures offers certain advantages over the use of modal properties. Lyon (1995) and Schultz et al. (1996) have promoted the use of measured frequency response functions for structural diagnostics.

Chen et al. (2011) applied the frequency domain technique to determine the Total Measurable Fault-Information-based Residual for fault detection in dynamic systems. A practical DC motor example, with a proportional–integral–derivative (PID) controller, was used to demonstrate the effectiveness of their method.

Prasannamoorthy and Devarajan (2010) applied the frequency domain technique for fault diagnosis in an analog-circuits software and hardware implementation. In both these cases, the signatures were extracted from the frequency response of the circuit and were found to be successful for the classification of faults.

Yu and Chao (2010) applied frequency domain data for fault diagnosis in squirrel cage induction motors. It was found that the method was successful in identifying fault characteristics.

Yeh et al. (2010) successfully applied frequency domain data for the detection of faults by using both control and output error signals, while Nandi et al. (2009) applied frequency domain data for the detection of faults in induction motors and Rishvanth et al. (2009) applied frequency domain data for short distance fault detection in optical fibers and integrated optical devices.

1.4.4 Time-Frequency Data

Some types of fault, such as cracks caused by fatigue failures, cause linear structures to become non-linear. In these cases, techniques such as linear finite element analysis and modal analysis cannot be applied and non-linear procedures are required (Ewins 1995; Marwala 2001). Non-linear structures give vibration data that are non-stationary. A non-stationary signal is one whose frequency components change as a function of time.

Illustrations of non-stationary signal include noise and vibration from an accelerating train. In order to analyze the non-stationary signal, the use of a Fast Fourier Transform (FFT) technique, which only displays the frequency components of the signal and is satisfactory for analyzing stationary signals, is not adequate here. As a result, time-frequency approaches that simultaneously show the time and frequency components of the signals are required. Some of the time-frequency approaches that have been used for fault identification are: the Short-Time Fourier Transform (STFT) (Newland 1993), the Wavelet Transform (WT) (Daubechies 1987), and the Wigner-Ville Distribution (WVD) (Wigner 1932).

Fundamentally the STFT transforms a small time window into a frequency domain. The time window is shifted to a new position and the Fourier transform is recurred. By doing so, a time-frequency spectrum is attained. If the time window is short, then the time-domain resolution becomes better and the frequency resolution becomes worse. Alternatively, if the time window is long, then the frequency-domain resolution becomes better and the time resolution becomes worse. Consequently, the time-frequency spectrum acquired from the STFT is limited in that any increase in the frequency resolution is at the cost of the time resolution. This drawback describes a principle called the Uncertainty Principle, which is analogous to Heisenberg's Uncertainty Principle (Wheeler and Zurek 1983), and in the current framework of signal processing may be assumed to be the result of producing a linear representation of a possibly non-linear signal. The STFT is said to be *linear*, as when computing it, the integral comprises a single, linear function of the signal and it is said to be *time-invariant* since the time shifted type of the signal results only in the time shifting of the time-frequency representation. In addition, the STFT is optimal for signals with a linearly increasing phase.

The WVD was established by Wigner (1932) in the framework of quantum mechanics and was brought to signal processing by Ville (1948). The WVD is based on the calculation of a correlation of a signal with itself (autocorrelation) to give an energy density. The Fourier transform of the calculated energy density gives the WVD. The WVD is understood to be bilinear because it uses two linear functions of the signal being analyzed, as opposed to one for the STFT, when calculating it. It affords an optimal representation of linear frequency modulation signals such as in a stationary frequency situation. The gains of the WVD are that it is optimized in both the time and frequency domain and that non-stationary signals display reduced distortion. The shortcomings of the WVD are that it does not account for the local behavior of the data at a given time and presents cross-terms when the signal being analyzed has many frequency components. The other difficulty, as described by Cohen (1989), is that this distribution spreads noise. It has been revealed that if there is noise present in a small segment of a signal, it is seen again within the WVD spectrum and this is related to the interference caused by cross-terms. The other problem with the WVD is that negative amplitude values may be attained in the results and this is physically irrelevant, making the results obtained from the WVD challenging to understand.

The WT decomposes the signal into a series of basis functions known as wavelets situated at different locations in the time axis in the same manner that the Fourier transform decomposes the signal into harmonic components. A given wavelet decays to zero at a distance away from its center. Local features of a signal can be recognized from the scale, which is similar to frequency, and the position in the time axis of the wavelets into which it is decomposed. A wavelet analysis allows the building of orthonormal bases with good time-frequency resolution.

Wavelets have the benefit in that they can identify local features of a signal from the frequency and the position in the time axis of the wavelets while the WVD does not actually describe the character of a signal at a given time. It gives an equal degree of significance to the far away times and the near times, making it non-local. The drawback of the wavelet method is that frequency is logarithmically scaled and, as a consequence, a low resolution is achieved at high frequencies (Barschdorf and Femmer 1995).

Surace and Ruotolo (1994) applied complex Morlet WTs to identify faults in a simulated cantilevered beam. The researchers found that for a fault simulated by a reduction of 20–45% in the beam's thickness, the amplitude of the WTs exhibited modulations that were consistent with the opening and closing of the crack.

Prime and Shevitz (1996) studied experimental data from a cantilevered beam with fatigue cracks of various magnitudes and observed that the 'harmonic mode shapes' are more sensitive to crack depth and location than are conventional mode shapes. The harmonic mode shapes were calculated using the magnitudes of harmonic peaks in the cross-power spectra. The researchers observed that the Wigner-Ville transforms were more sensitive to non-linearity than were the Fourier transforms.

Treetrong (2011b) applied a time-frequency analysis for the fault prediction of an induction motor and found that the presented technique provided a good accuracy in fault prediction and fault level quantification. Qian et al. (2010) successfully applied the STFT for the fault diagnosis of an air-lift compressor for an offshore oil and gas platform. Li et al. (2010) applied a Hough transform, which was adopted to analyze the Wigner-Ville time-frequency distribution, for rolling bearing fault diagnosis. Pattern recognition techniques were applied and the results showed that the Hough transform of Wigner-Ville time-frequency image can successfully classify the rolling bearing faults.

Borghetti et al. (2010) successfully applied time-frequency wavelet analysis for the fault location of distribution networks. Ma et al. (2009) successfully applied wavelet analysis to detect oil-film instability faults in rotor systems and Wei et al. (2009) successfully applied neural network modeling and wavelet processed data for the fault diagnosis of aircraft power plants.

Ma et al. (2010) also successfully applied a wavelet time-frequency feature analysis of oil-film instability faults in a rotor system. Vibration signals with two different types of parameters were gathered by changing the thickness of disc and shaft length, which was analyzed using a wavelet transform.

One weakness of time-frequency methods is that there are many types, including WT, WVD and STFT, and there is no methodical technique to select the most suitable kind for fault identification. Nevertheless, comparative studies have shown that wavelet transforms are better suited for the fault detection problem than are the WVD and STFT. Nonetheless, time-frequency methods have the following advantages: one measurement provides abundant data; and they are effective in identifying faults that result in the loss of linearity of a structure.

1.5 Strategies Used for Condition Monitoring

This section explains the most common strategies that have been applied for condition monitoring in structures using vibration data in many domains. The three strategies considered are correlation based models, finite element updating techniques and computational intelligence techniques.

1.5.1 Correlation Based Methods

Correlation based techniques apply vibration data in the frequency or modal domains to identify faults. They are computationally cheaper to apply than approaches that use complicated mathematical models. The modal assurance criterion (MAC) (Allemang and Brown 1982) and the coordinate modal assurance criterion (COMAC) (Lieven and Ewins 1988), are measures of correlation between mode shapes, and have been used to identify faults in structures (West 1984; Fox 1992; Kim et al. 1992; Salawu and Williams 1994; Lam et al. 1995; Marwala 2001). The curvature was calculated using the central difference approximation technique. Messina et al. (1998) introduced the multiple fault location assurance criterion, which applied the correlation between the natural frequencies from faulty and healthy structures to identify the location and size of faults.

Maia et al. (1997) applied the frequency-response-function-curvature technique which is the difference between curvatures of faulty and healthy structures to identify faults. The response-function-quotient technique used quotients between the frequency response function at different locations for fault detection (Maia et al. 1999). Gawronski and Sawicki (2000) used modal norms to successfully identify faults in structures. The modal norms were estimated from the natural frequencies, modal damping and modal displacements at the actuator and sensor locations of healthy and faulty structures. Worden et al. (2000) applied outlier analysis to detect fault on various simulated structures and a carbon fiber plate by comparing the deviation of a transmissibility-function signal from what is considered normal.

Rolo-Naranjo and Montesino-Otero (2005) applied a correlation dimension approximation for the on-line condition monitoring of large rotating machinery. This technique was based on a systemic analysis of the second derivative of the correlation integral obtained from the Grassberger and Procaccia algorithm. The results revealed the applicability of the technique in vibration-signal analysis based condition monitoring.

1.5.2 Finite Element Updating Techniques

The finite element model updating technique has been used to identify faults on structures (Friswell and Mottershead 1995; Maia and Silva 1997; Marwala 2010). When implementing the finite element updating techniques for identification, it is assumed that the finite element model is a true dynamic representation of the structure. This means that changing any physical parameter of an element in the finite element model is equivalent to introducing a fault in that region.

There are two techniques used in finite element updating (Friswell and Mottershead 1995): direct techniques and iterative methods. Direct methods, which use the modal properties, are computationally efficient to implement and reproduce the measured modal data exactly. They do not take into account the physical parameters that are updated.

Iterative procedures use changes in physical parameters to update finite element models and produce models that are physically realistic.

Finite element updating approaches are implemented by minimizing the distance between analytical and measured data. The difference between the updated systems matrices and original matrices identifies the presence, location and extent of faults. One way of implementing this procedure is to formulate the objective function to be minimized and choose an optimization routine (Marwala 2010). Some of the optimization methods that have been used in the past are particle swarm optimization, genetic algorithm, simulated annealing and a hybrid of a number of techniques (Marwala 2010). These procedures are classified as iterative because they are implemented by iteratively modifying the relevant physical parameters of the model until the error is minimized.

The approaches described in this subsection are computationally expensive because they require an optimization method. In addition, it is difficult to find a global minimum through the optimization technique, due to the multiple stationary points, which are caused by its non-unique nature (Janter and Sas 1990). Techniques such as the use of genetic algorithms and multiple starting design variables have been applied in the past to increase the probability of finding the global minimum (Mares and Surace 1996; Larson and Zimmerman 1993; Dunn 1998).

Sensitivity based approaches assume that experimental data are perturbations of design data about the original finite element model. Due to this assumption, experimental data must be close to the finite element data for these approaches to be effective. This formulation only works if the structural modification is small. These approaches are based on the estimation of the derivatives of either the modal properties or the frequency response functions. Many techniques have been developed to estimate the derivative of the modal properties and frequency response functions. Norris and Meirovitch (1989), Haug and Choi (1984) and Chen and Garba (1980) presented other procedures of computing the derivatives of the modal properties to ascertain parameter changes. They used orthogonal relations for the mass and stiffness matrices to compute the derivatives of the natural frequencies and mode shapes with respect to parameter changes. Ben-Haim and Prells (1993) proposed selective FRF sensitivity to uncouple the finite element updating problem. Lin et al. (1995) improved the modal sensitivity technique by ensuring that these approaches were applicable to large magnitude faults.

Hemez (1993) proposed a technique that assesses the sensitivity at an element level. The advantage of this technique is its ability to identify local errors. In addition, it is computationally efficient. Alvin (1996) modified this technique to improve the convergence rate by using a more realistic error indicator and by incorporating statistical confidence measurements for both the initial model parameters and the measured data.

Eigenstructure assignment methods are based on control system theory. The structure under investigation is forced to respond in a predetermined manner. During fault detection, the desired eigenstructure is the one that is measured in the test. Zimmerman and Kaouk (1992) applied these techniques to identify the elastic modulus of a cantilevered beam using measured modal data. Schultz et al. (1996) improved this method by using measured Frequency Response Functions (FRFs).

The one limitation of the methods outlined in this section is that the number of sensor locations is less than the number of degrees of freedom in the finite element model. This is especially problematic since it renders the integration of the experimental data and finite element model – the very basis of finite element updating fault identification methods – difficult. To compensate for this limitation, the mode shapes and FRFs are either expanded to the size of the finite element model or the mass and stiffness matrices of the finite element model are reduced to the size of the measured data. Among the reduction methods that have been applied are the static reduction (Guyan 1965; Marwala 2001), dynamic reduction (Paz 1984; Marwala 2001), improved reduced system and system-equivalent-reduction-process (O'Callahan et al. 1989; Marwala 2001). Techniques that expand the mass and stiffness matrices have also been employed (Gysin 1990; Imregun and Ewins 1993; Marwala 2001).

It has been shown that finite element updating techniques have numerous limitations. Most importantly, they rely on an accurate finite element model, which may not be available. Even if the model is available, the problem of the non-uniqueness of the updated model makes the problem of fault identification using finite element updating non-unique.

Purbolaksono et al. (2009) successfully applied finite element modeling for the supplemental condition monitoring of a water-tube boiler. The technique used empirical formula for approximating the scale thickness developed on the inner surface of the tube over period of time.

1.5.3 Computational Intelligence Methods

In recent times, there has been increased interest in applying computational artificial neural networks to identify faults in structures. Neural networks can estimate functions of arbitrary complexity using given data. Supervised neural networks are used to represent a mapping from an input vector onto an output vector, while unsupervised networks are used to classify the data without prior knowledge of the classes involved. The most common neural network architecture is the Multi-Layer Perceptron (MLP), trained using the back-propagation technique (Bishop 1995). An alternative network is the radial basis function (RBF) (Bishop 1995).

Kudva et al. (1991) used MLP neural networks to identify faults on a plate. The inputs to the neural network were the readings from a strain gauge, obtained by applying a static uniaxial load to the structure, while the output was the location and size of a hole. The fault was modeled by cutting holes of diameters that varied from 12.7 to 63.5 mm. The authors found that the neural network could predict the error location without failure, although difficulty was experienced in predicting the size of a hole. In cases where the neural network successfully identified the size of a hole, there was approximately a 50% error.

Wu et al. (1992) used an MLP neural network to identify faults in a model of a three-story building. Faults were modeled by reducing the member stiffness by between 50% and 75%. The input to the neural network was the Fourier transform of the acceleration data, while the output was the level of fault in each member. The network was able to diagnose faults within an accuracy of 25%.

Leath and Zimmerman (1993) applied the MLP to identify faults on a four-element cantilevered beam, which was modeled by reducing the Young's modulus by up to 95%. The inputs to the neural network were the first two natural frequencies and the output was Young's modulus. The neural network could identify faults to within an accuracy of 35%.

Worden et al. (1993) used an MLP neural network to identify faults in a twenty-member structure, which was modeled by removing each member. The input to the neural network was the strain in twelve members. The network was trained using data from the finite element model. When applied to experimental data, the network usually could detect the location of the fault. Atalla (1996) trained a RBF neural network using Frequency Response Functions in order to identify faults in structures.

Widodo and Yang (2007) reviewed the successful application of Support Vector Machines in machine condition monitoring and fault diagnosis while Bouhouche et al. (2010) applied online Support Vector Machines and fuzzy reasoning for the condition monitoring of the hot rolling process.

Aliustaoglu et al. (2009) successfully applied a fuzzy system to tool wear condition monitoring while Lau and Dwight (2011) successfully applied a fuzzy-based decision support model for the condition monitoring of water pipelines. Wong et al. (2010) successfully applied log-polar mapping, quaternion correlation and max-product fuzzy neural network for a thermal condition monitoring system while Weidl et al. (2005) applied object-oriented Bayesian networks for the condition monitoring, root cause analysis and decision support of complex continuous processes.

The finite element updating methods discussed in Sect. 1.4.2 require the availability of an accurate finite element model to perform fault identification, which may not be available. Methods in Sect. 1.4.2 avoid the need for a finite element model but can mostly only detect faults and do not seem to be able to locate and quantify faults well. The implementation of computational intelligence methods does not necessarily require the availability of a finite element model but requires that the vibration data be available to train the network and can detect, locate and quantify faults.

1.6 Summary of the Book

In Chap. 2, the data gathering and processing methods that are used for condition monitoring in this book are reviewed. Different data gathering techniques for condition monitoring and the essential elements of data gathering within the context of condition monitoring are outlined. These include issues such as data type, measuring instruments, sampling frequencies, leakages and measurement errors. In particular, Fourier transform, the modal domain data, pseudo-modal energies, wavelet transform and Mel-frequency data are reviewed. In addition, the method for data visualization reviewed is the principal component analysis.

In Chap. 3, neural networks methods are introduced for condition monitoring. In particular, the Multi-Layer Perceptron (MLP) neural network is introduced. It is trained using the maximum-likelihood technique. The MLP is then applied for fault identification in a population of cylindrical shells.

In Chap. 4, Bayesian neural networks methods are introduced for condition monitoring. In particular, the Multi-Layer Perceptron (MLP) neural network, trained using a hybrid Monte Carlo simulation is introduced. The MLP is then applied for fault identification in a population of cylindrical shells.

In Chap. 5, a committee of networks is introduced. This committee is made of three Multi-Layer Perceptrons one with the wavelet data as input, the other one with modal properties as inputs and the third with pseudo-modal energies as inputs. It is mathematically and empirically demonstrated that the committee is better than the individual techniques.

In Chap. 6, Gaussian mixture models and hidden Markov models are applied for condition monitoring in mechanical structures. These methods are described, implemented, and compared.

In Chap. 7, fuzzy system methods are applied for condition monitoring. They are fuzzy logic and the fuzzy ARTMAP and are described implemented for the condition monitoring.

In Chap. 8, rough systems are explained and applied for the condition monitoring of transformer bushings, while in Chap. 9, a method for fault classification in mechanical systems in the presence of missing data entries is introduced. The method constructed is based on auto-associative neural networks where the network is trained to recall the input data through some nonlinear neural network mapping from the trained network with an error equation with missing inputs as design variables. A genetic algorithm is used to solve for the missing input values. The presented method is tested on a fault classification problem in a population of cylindrical shells.

In Chap. 10, condition monitoring using support vector machine and extension neural networks is introduced. The theories of the support vector machine and extension neural networks are described, implemented and compared.

In Chap. 11, condition monitoring using incremental learning is presented. The ability of a classifier to take on new information and classes by evolving the classifier without it having to be fully retrained is known as *incremental learning*.

In the chapter a Learn++ incremental learning algorithms is applied for the condition monitoring in transformer bushings.

In Chap. 12, the condition monitoring methods described in this book are compared and then conclusions are drawn. In addition, future and emerging areas in condition monitoring are identified and emerging opportunities are highlighted.

References

Abdel-Latief AN, Abdel-Gawad AF, Ishak AA, Mandour ME (2003) Fault type classification in power distribution feeders utilizing statistical functions and neural networks. In: Proceedings of the universities power engineering conference, Thessaloniki, Greece, pp 614–617

Al-Habaibeh A, Zorriassatine F, Gindy N (2002) Comprehensive experimental evaluation of a systematic approach for cost effective and rapid design of condition monitoring systems using Taguchi's method. J Mater Process Technol 124:372–383

Aliustaoglu C, Ertunc HM, Ocak H (2009) Tool wear condition monitoring using a sensor fusion model based on fuzzy inference system. Mech Syst Signal Process 23:539–546

Allemang RJ, Brown DL (1982) A correlation coefficient for modal vector analysis. In: Proceedings of the 1st international modal analysis conference, Orlando, Florida, pp 1–18

Alvin KF (1996) Finite element model updating via bayesian estimation and minimisation of dynamic residuals. In: Proceedings of the 14th international modal analysis conference, Dearbon, Michigan, pp 428–431

Andrade FA, Esat I, Badi MNM (2001) A new approach to time-domain vibration condition monitoring: gear tooth fatigue crack detection and identification by the Kolmogorov-Smirnov test. J Sound Vib 240:909–919

Atalla MJ (1996) Model updating using neural networks. PhD thesis, Virginia Polytechnic Institute and State University, Blacksburg, Virginia

Barschdorf D, Femmer U (1995) Signal processing and pattern recognition methods for biomedical sound analysis. In: Proceedings of the 2nd international symposium of acoustical and vibratory surveillance methods and diagnostic techniques, Paris, France, pp 279–290

Bartelmus W, Zimroz R (2009) A new feature for monitoring the condition of gearboxes in non-stationary operating conditions. Mech Syst Signal Process 23:1528–1534

Ben-Haim Y, Prells U (1993) Selective sensitivity in the frequency domain, Part I: Theory. Mech Syst Signal Process 7:461–475

Bishop CM (1995) Neural networks for pattern recognition. Oxford University Press, Oxford

Booth C, McDonald JR (1998) The use of artificial neural networks for condition monitoring of electrical power transformers. Neurocomputing 23:97–109

Borghetti A, Bosetti M, Nucci CA, Paolone M, Abur A (2010) Integrated use of time-frequency wavelet decompositions for fault location in distribution networks: theory and experimental validation. IEEE Trans Power Deliv 25:3139–3146

Bouhouche S, Yazid LL, Hocine S, Bast J (2010) Evaluation using online support-vector-machines and fuzzy reasoning. Application to condition monitoring of speeds rolling process. Control Eng Pract 18:1060–1068

Bouhouche S, Yahi M, Bast J (2011) Combined use of principal component analysis and self organisation map for condition monitoring in pickling process. Appl Soft Comput J 11:3075–3082

Butler S, Ringwood J (2010) Particle filters for remaining useful life estimation of abatement equipment used in semiconductor manufacturing. In: Proceedings of the conference on control and fault-tolerant systems, Nice, France, pp 436–441

Cawley P, Adams RD (1979) The location of defects from measurements of natural frequencies. J Strain Anal 14:49–57

Chen JC, Garba JA (1980) Analytical model improvement using modal test results. Am Inst Aeronaut Astronaut J 18:684–690

Chen W, Yeh CP, Yang H (2011) ToMFIR-based fault detection approach in frequency domain. J Syst Eng Electron 22:33–37

Cohen L (1989) Time-frequency distributions – a review. In: Proceedings of the IEEE, pp 941–981

D'Ambrogio W, Zobel PB (1994) Damage detection in truss structures using a direct updating technique. In: Proceedings of the 19th international seminar for modal analysis, Leuvel, Belgium, pp 657–667

Daubechies I (1987) Orthogonal bases of wavelets with finite support connection with discrete filters. In: Proceedings of the international conference on wavelets, Marseille, France, pp 38–66

Doebling SW, Farrar CR, Prime MB, Shevitz DW (1996) Damage identification and health monitoring of structural and mechanical systems from changes in their vibration characteristics: a literature review. Los Alamos National Laboratory report LA-13070-MS, Los Alamos

Dunn SA (1998) The use of genetic algorithms and stochastic hill-climbing in dynamic finite-element model identification. Comput Struct 66:489–497

Elangovan M, Ramachandran KI, Sugumaran V (2010) Studies on bayes classifier for condition monitoring of single point carbide tipped tool based on statistical and histogram features. Expert Syst Appl 37:2059–2065

Ewins DJ (1995) Modal testing: theory and practice. Research Studies Press, Letchworth

Farrar CR, Baker WE, Bell TM, Cone KM, Darling TW, Duffey TA, Eklund A, Migliori A (1994) Dynamic characteristics and damage detection in the I-40 bridge over the Rio Grande. Los Alamos National Laboratory report LA-12767-MS, Los Alamos

Fox CHJ (1992) The location of defects in structures: a comparison of the use of natural frequency and mode shape data. In: Proceedings of the 10th international modal analysis conference, San Diego, California, pp 522–528

Friswell MI, Mottershead JE (1995) Finite element model updating in structural dynamics. Kluwer Academic Publishers Group, Dordrecht

Garcia-Escudero LA, Duque-Perez O, Morinigo-Sotelo D, Perez-Alonso M (2011) Robust condition monitoring for early detection of broken rotor bars in induction motors. Expert Syst Appl 38:2653–2660

Garvey DR, Baumann J, Lehr J, Hughes B, Hines JW (2009) Pattern recognition-based remaining useful life estimation of bottomhole assembly tools. In: Proceedings of the SPE/IADC drilling conference, Bahrain, pp 82–89

Gawronski W, Sawicki JT (2000) Structural damage detection using modal norms. J Sound Vib 229:194–198

Gedafa DS, Hossain M, Miller R, Van T (2010) Estimation of remaining service life of flexible pavements from surface deflections. J Transp Eng 136:342–352

Gunal S, Ece DG, Gerek ON (2009) Induction machine condition monitoring using notch-filtered motor current. Mech Syst Signal Process 23:2658–2670

Guyan RJ (1965) Reduction of stiffness and mass matrices. Am Inst Aeronaut Astronaut J 3:380

Gysin H (1990) Comparison of expansion methods for FE model localization. In: Proceedings of the 8th international modal analysis conference, Kissimmee, Florida, pp 195–204

Haroon M, Adams DE (2007) Time and frequency domain nonlinear system characterization for mechanical fault identification. Nonlinear Dyn 50:387–408

Haug EF, Choi KK (1984) Structural design sensitivity with generalized global stiffness and mass matrices. Am Inst Aeronaut Astronaut J 22:1299–1303

He Q, Yan R, Kong F, Du R (2009) Machine condition monitoring using principal component representations. Mech Syst Signal Process 23:446–466

He ZY, Li XP, Wu LY, Xia LL, Qian QQ (2010) A new single ended fault location method using travelling wave natural frequencies – Part 2: Problems concerned in implementation. In: Proceedings of the IEEE PES general meeting, Minneapolis, USA, pp 1–6

Hemez FM (1993) Theoretical and experimental correlation between finite element models and modal tests in the context of large flexible structures. PhD thesis, University of Colorado

Huang SF, Huang H, Wang XG, Wang LC (2009) Analysis on natural frequencies of ultra-high voltage transmission lines in short circuit fault. High Volt Eng 35:2059–2065

Hussain S, Gabbar HA (2011) A novel method for real time gear fault detection based on pulse shape analysis. Mech Syst Signal Process 25:1287–1298

Imregun M, Ewins DJ (1993) An investigation into mode shape expansion techniques. In: Proceedings of the 11th international modal analysis conference, Kissimmee, Florida, pp 168–175

Imregun M, Visser WJ, Ewins DJ (1995) Finite element model updating using frequency response function data – Part I: Theory and initial investigation. Mech Syst Signal Process 9:187–202

Jain A, Thoke AS, Koley E, Patel RN (2009) Fault classification and fault distance location of double circuit transmission lines for phase to phase faults using only one terminal data. In: Proceedings of the international conference on power systems, Kharagpur, India, pp 1–6

Janter T, Sas P (1990) Uniqueness aspects of model-updating procedure. Am Inst Aeronaut Astronaut J 28:538–543

Jayabharata Reddy M, Mohanta DK (2011) A modular approach for classification and location of arcing and non-arcing faults on transmission lines. Int J Energy Technol Policy 7:309–324

Kazemi S, Fooladi A, Rahai AR (2010) Implementation of the modal flexibility variation to fault identification in thin plates. Acta Astronaut 66:414–426

Kazemi S, Rahai AR, Daneshmand F, Fooladi A (2011) Implementation of modal flexibility variation method and genetically trained ANNs in fault identification. Ocean Eng 38:774–781

Khorashadi-Zadeh H, Li Z (2008) Fault classification and fault location using ANN for medium voltage cables: design and implementation. Intell Autom Soft Comput 14:479–489

Khosravi A, Llobet JA (2007) A hybrid method for fault detection and modelling using modal intervals and ANFIS. In: Proceedings of the American control conference, New York, USA, pp 3003–3008

Kim JH, Jeon HS, Lee SW (1992) Application of modal assurance criteria for detecting and locating structural faults. In: Proceedings of the 10th international modal analysis conference, San Diego, California, pp 536–540

Kim S, Kim S, Choi H (2010a) Remaining life estimation of a level luffing crane component by computer simulation. J Korean Inst Met Mater 48:489–497

Kim YS, Lee DH, Kim SK (2010b) Fault classification for rotating machinery using support vector machines with optimal features corresponding to each fault type. Trans Korean Soc Mech Eng 34:1681–1689

Kudva J, Munir N, Tan P (1991) Damage detection in smart structures using neural networks and finite element analysis. In: Proceedings of the ADPA/AIAA/ASME/SPIE conference on active materials and adaptive structures, Alexandria, Virginia, pp 559–562

Lai JY, Young KF (1995) Dynamics of graphite/epoxy composite under delamination fracture and environmental effects. J Comput Struct 30:25–32

Lam HF, Ko JM, Wong CW (1995) Detection of damage location based on sensitivity analysis. In: Proceedings of the 13th international modal analysis conference, Nashville, Teneessee, pp 1499–1505

Larson CB, Zimmerman DC (1993) Structural model refinement using a genetic algorithm approach. In: Proceedings of the 11th international modal analysis conference, Kissimmee, Florida, pp 1095–1101

Lau HCW, Dwight RA (2011) A fuzzy-based decision support model for engineering asset condition monitoring – a case study of examination of water pipelines. Expert Syst Appl 38:13342–13350

Leath WJ, Zimmerman DC (1993) Analysis of neural network supervised training with application to structural damage detection. Damage and control of large structures. In: Proceedings of the 9th Virginia Polytechnic Institute and State University symposium, Blacksburg, Virginia, pp 583–594

Lee BT, Sun CT, Liu D (1987) An assessment of damping measurement in the evaluation of integrity of composite beams. J Reinf Plast Compos 6:114–125

Li H, Zhang Z, Guo Z, Zou S, Wang F (2010) Rolling bearing fault diagnosis using hough transform of time-frequency image. J Vib Meas Diagn 30:634–637

Lieven NAJ, Ewins DJ (1988) Spatial correlation of mode shapes, the co-ordinate modal assurance criterion. In: Proceedings of the 6th international modal analysis conference, Kissimmee, Florida, pp 690–695

Lifshitz JM, Rotem A (1969) Determination of reinforcement unbonding of composites by a vibration technique. J Compos Mater 3:412–423

Liguo Z, Yutian W, Sheng Z, Guangpu H (2009) The fault diagnosis of machine based on modal analysis. In: Proceedings of the international conference on measuring technology and mechatronics automation, Hunan, China, pp 738–741

Lin RM, Lim MK, Du H (1995) Improved inverse eigensensitivity method for structural analytical model updating. J Vib Acoust 117:192–198

Lin S, He Z, Zang T, Qian Q (2010) Novel approach of fault type classification in transmission lines based on rough membership neural networks. Proc Chin Soc Electr Eng 30:72–79

Loutas TH, Roulias D, Pauly E, Kostopoulos V (2011) The combined Use of vibration, acoustic emission and Oil debris on-line monitoring towards a more effective condition monitoring of rotating machinery. Mech Syst Signal Process 25:1339–1352

Luo GY, Osypiw D, Irle M (2000) Real-time condition monitoring by significant and natural frequencies analysis of vibration signal with wavelet filter and autocorrelation enhancement. J Sound Vib 236:413–430

Lyon R (1995) Structural diagnostics using vibration transfer functions. Sound Vib 29:28–31

Ma H, Zhang L, Li H, Xie W (2007a) The application of modal analysis in fault diagnosis of AC motor. In: Proceedings of the international conference on condition monitoring and diagnosis, pp 217–220

Ma H, Li H, Xie W, Chen F (2007b) Vibration research on winding faults of induction motor based on experiment modal analysis method. In: Proceedings of the 8th international power engineering conference, Guangdon, China, pp 366–370

Ma H, Sun W, Ren Z, Wen B (2009) Feature analysis of oil-film instability fault based on time-frequency methods in rotor systems. In: Proceedings of the 2nd international conference on intelligent computation technology and automation, Hunan, China, pp 541–544

Ma H, Li CF, Xuan GJ, Wen BC (2010) Time-frequency feature analysis of oil-film instability fault in a rotor system. J Vib Shock 29:193–195+198

Maia NMM, Silva JMM (1997) Theoretical and experimental modal analysis. Research Studies Press, Letchworth

Maia NMM, Silva JMM, Sampaio RPC (1997) Localization of damage using curvature of the frequency-response-functions. In: Proceedings of the 15th international modal analysis conference, Orlando, Florida, pp 942–946

Maia NMM, Silva JMM, Sampaio RPC (1999) On the use of frequency response functions for damage detection. In: Proceedings of the 2nd international conference on identification in engineering system, Kissimmee, Florida, pp 460–471

Mares C, Surace C (1996) An application of genetic algorithms to identify damage in elastic structures. J Sound Vib 195:195–215

Marwala T (1997) Multi-criteria method for determining damage on structures. Master's thesis, University of Pretoria, Pretoria, South Africa

Marwala T (2001) Fault identification using neural networks and vibration data. PhD thesis, University of Cambridge, Cambridge, UK

Marwala T (2010) Finite element model updating using computational intelligence techniques. Springer, London

Mayes RL (1992) Error localization using mode shapes – an application to a two link robot arm. In: Proceedings of the 10th international modal analysis conference, San Diego, California, pp 886–891

Messina A, Jones IA, Williams EJ (1996) Damage detection and localisation using natural frequency changes. In: Proceedings of the 1st international conference on identification in engineering system, Swansea, Wales, pp 67–76

Messina A, Williams EJ, Contursi T (1998) Structural damage detection by a sensitivity and statistical-based method. J Sound Vib 216:791–808
Migliori A, Bell TM, Dixon RD, Strong R (1993) Resonant ultrasound nondestructive inspection. Los Alamos National Laboratory report LS-UR-93-225, Los Alamos, Las Vegas, Nevada
Miya WS, Mpanza LJ, Nelwamondo FV, Marwala T (2008) Condition monitoring of oil-impregnated paper bushings using extension neural network, Gaussian mixture models and hidden Markov models. In: Proceedings of the IEEE international conference on man, systems, and cybernetics, Singapore, pp 1954–1959
Murthy VS, Mohanta DK, Gupta S (2011) Video surveillance-based insulator condition monitoring analysis for substation monitoring system (SMS). Int J Inf Commun Technol 3:11–31
Nandi S, Ilamparithi T, Lee SB, Hyun D (2009) Pole pair and rotor slot number independent frequency domain based detection of eccentricity faults in induction machines using a semi on-line technique. In: Proceedings of the IEEE international symposium on diagnostics for electric machines, power electronics and drives, Cargese, France, pp 1–7
Nelwamondo FV, Marwala T (2007) Techniques for handling missing data: applications to online condition monitoring. Int J Innov Comput Inf Control 4:1507–1526
Newland DE (1993) An introduction to random vibration, spectral and wavelet analysis. Longman/Harlow/Wiley, New York
Norris MA, Meirovitch L (1989) On the problem of modelling for parameter identification in distributed structures. Int J Numer Methods Eng 28:2451–2463
O'Callahan JC, Avitabile P, Riemer R (1989) System equivalent reduction expansion process. In: Proceedings of the 7th international modal analysis conference, Las Vegas, Nevada, pp 17–21
Pandurangaiah D, Reddy CC, Govindan TP, Mandlik M, Ramu TS (2008) Estimation of remaining life of power transformers. In: Proceedings of the IEEE international symposium on electrical insulation, Vancouver, Canada, pp 243–246
Park S, Kim JW, Lee C, Park SK (2011) Impedance-based wireless debonding condition monitoring of CFRP laminated concrete structures. NDT E Int 44:232–238
Paz M (1984) Dynamic condensation. Am Inst Aeronaut Astronaut J 22:724–727
Pedregal DJ, Carnero MC (2006) State space models for condition monitoring: a case study. Reliab Eng Syst Saf 91:171–180
Prasannamoorthy V, Devarajan N (2010) Frequency domain technique for fault diagnosis in analog circuits – software and hardware implementation. J Theor Appl Inf Technol 22:107–119
Prime MB, Shevitz DW (1996) Damage detection of street frame by modal testing. In: Proceedings of the 11th international modal analysis conference, Kissimmee, Florida, pp 1437–1443
Purbolaksono J, Khinani A, Ali AA, Rashid AZ, Nordin NF (2009) Iterative technique and finite element simulation for supplemental condition monitoring of water-tube boiler. Simul Model Pract Theory 17:897–910
Qian S, Jiao W, Hu H (2010) Time-frequency analysis and fault diagnosis of air-lift compressor for an offshore oil and gas platform. In: Proceedings of the 29th Chinese control conference, Beijing, China, pp 2977–2980
Riml S, Piontke A, Larcher L, Kompatscher P (2010) Quantification of faults resulting from disregard of standardised facial photography. J Plast Reconstr Aesthet Surg 64:898–901
Rishvanth KP, Rai S, Kumar S, Sudheer SK, Raina JP (2009) Design and simulation of optical frequency domain reflectometer for short distance fault detection in optical fibers and integrated optical devices using Ptolemy-II. In: Proceedings of the international conference on ultra modern telecommunications and workshops, St. Petersburg, Russia, pp 1–3
Rolo-Naranjo A, Montesino-Otero ME (2005) A method for the correlation dimension estimation for on-line condition monitoring of large rotating machinery. Mech Syst Signal Process 19:939–954
Salawu OS (1995) Non-destructive assessment of structures using integrity index method applied to a concrete highway bridge. Insight 37:875–878
Salawu OS, Williams C (1994) Damage location using vibration mode shapes. In: Proceedings of the 11th international modal analysis conference, Kissimmee, Florida, pp 933–939

Schultz MJ, Warwick DN (1971) Vibration response: a non-destructive test for fatigue crack damage in filament-reinforced composites. J Compos Mater 5:394–404
Schultz MJ, Pai PF, Abdelnaser AS (1996) Frequency response function assignment technique for structural damage identification. In: Proceedings of the 14th international modal analysis conference, Dearborn, Michigan, pp 105–111
Sestieri A, D'Ambrogio W (1989) Why be modal: how to avoid the use of modes in the modification of vibrating systems. In: Proceedings of the 7th international modal analysis conference, Las Vegas, Nevada, pp 25–30
Sinha JK (2009) Recent trends in fault quantification in rotating machines. Adv Vib Eng 8:79–85
Surace C, Ruotolo R (1994) Crack detection of a beam using the wavelet transform. In: Proceedings of the 12th international modal analysis conference, Honolulu, Hawaii, pp 1141–1148
Tao B, Zhu L, Ding H, Xiong Y (2007) An alternative time-domain index for condition monitoring of rolling element bearings – a comparison study. Reliab Eng Syst Saf 92:660–670
Thai N, Yuan L (2011) Transmission line fault type classification based on novel features and neuro-fuzzy system. Electr Power Compon Syst 38:695–709
Tian Z, Jin T, Wu B, Ding F (2011) Condition based maintenance optimization for wind power generation systems under continuous monitoring. Renew Energy 36:1502–1509
Treetrong J (2011a) Fault prediction of induction motor based on time-frequency analysis. Appl Mech Mater 52–54:115–120
Treetrong J (2011b) The use of higher-order spectrum for fault quantification of industrial electric motors. Lect Notes Electr Eng 70:59–68
Vilakazi CB, Marwala T (2007) Incremental learning and its application to bushing condition monitoring. Lect Notes Comput Sci 4491:1241–1250
Vilakazi CB, Marwala T (2009) Computational intelligence approach to Bushing condition monitoring: incremental learning and its application. In: Intelligent engineering systems and computational cybernetics, pp 161–171. doi:10.1007/978/1-4020-8678-6-14
Vilakazi CB, Mautla RP, Moloto EM, Marwala T (2005) On-line Bushing condition monitoring. In: Proceedings of the 5th WSEAS/IASME international conference on electric power system, Steven's Point, Wisconsin, pp 406–411
Ville J (1948) Théorie et Applications de la Notion de Signal Analytique. Cables et Transm 2A:61–74
Wei L, Hua W, Pu H (2009) Neural network modeling of aircraft power plant and fault diagnosis method using time frequency analysis. In: Proceedings of the Chinese control and decision conference, pp 353–356
Weidl G, Madsen AL, Israelson S (2005) Applications of object-oriented Bayesian networks for condition monitoring, root cause analysis and decision support on operation of complex continuous processes. Comput Chem Eng 29:1996–2009
West WM (1984) Illustration of the use of modal assurance criterion to detect structural changes in an orbiter test specimen. In: Proceedings of air force conference on aircraft structural integrity, Palm Springs, California, pp 1–6
Wheeler JA, Zurek H (1983) Quantum theory and measurement. Princeton University Press, Princeton
Widodo A, Yang BS (2007) Support vector machine in machine condition monitoring and fault diagnosis. Mech Syst Signal Process 21:2560–2574
Wigner EP (1932) On quantum correction for thermodynamic equilibrium. Phys Rev 40:749–759
Wong WK, Loo CK, Lim WS, Tan PN (2010) Thermal condition monitoring system using log-polar mapping, quaternion correlation and Max-product fuzzy neural network classification. Neurocomputing 74:164–177
Worden KA, Ball A, Tomilinson G (1993) Neural networks for fault location. In: Proceedings of the 11th international modal analysis conference, Kissimmee, Florida, pp 47–54
Worden K, Manson G, Fieler NRJ (2000) Damage detection using outlier analysis. J Sound Vib 229:647–667
Wu X, Ghaboussi J, Garret JH (1992) Use of neural networks in detection of structural damage. Comput Struct 42:649–659

Xia L, He Z, Li X, Chen S (2010) A fault location method based on natural frequencies and empirical mode decomposition for mixed overhead-cable lines. Autom Electr Power Syst 34:67–73

Xie CL, Liu YK, Xia H (2009) Application of ant colony optimization in NPP classification fault location. Nucl Power Eng 30:108–112

Yadav SK, Tyagi K, Shah B, Kalra PK (2011) Audio signature-based condition monitoring of internal combustion engine using FFT and correlation approach. IEEE Trans Instrum Meas 60:1217–1226

Yanagawa T, Kaneko M, Iida I, Fujiwara R (2010) Development of new remaining life estimation method for main parts of hydro-turbine in hydro electric power station. In: Proceedings of the AIP conference, New York, USA, pp 834–839

Yeh CP, Yang HL, Chen W (2010) A fault detection approach using both control and output error signals in frequency domain. In: Proceedings of the IEEE/ASME international conference on mechatronic and embedded systems and applications, Suzhou, China, pp 341–344

Yu W, Chao S (2010) Fault diagnosis way based on RELAX algorithms in frequency domain for the squirrel cage induction motors. In: Proceedings of the international conference on computer intelligence and software engineering, Wuhan, China, pp 1–4

Zhang L, Huang AQ (2011) Model-based fault detection of hybrid fuel cell and photovoltaic direct current power sources. J Power Sour 196:5197–5204

Zhang Y, Suonan J (2010) Time domain fault location method based on UHV transmission line parameter identification using two terminal data. In: Proceedings of the Asia-Pacific power and energy engineering conference, Wuhan, China, pp 1–5

Zhou JH, Pang CK, Lewis FL, Zhong ZW (2011a) Dominant feature identification for industrial fault detection and isolation applications. Expert Syst Appl 38:10676–10684

Zhou Y, Tao T, Mei X, Jiang G, Sun N (2011b) Feed-axis gearbox condition monitoring using built-in position sensors and EEMD method. Robot Comput Integr Manuf 27:785–793

Zhu K, Wong YS, Hong GS (2009) Wavelet analysis of sensor signals for tool condition monitoring: a review and some new results. Int J Mach Tools Manuf 49:537–553

Zhu Y, Wang W, Tong S (2011) Fault detection and fault-tolerant control for a class of nonlinear system based on fuzzy logic system. ICIC Expr Lett 5:1597–1602

Zi Y, Chen X, He Z, Chen P (2005) Vibration based modal parameters identification and wear fault diagnosis using Laplace wavelet. Key Eng Mater 293–294:83–190

Zimmerman DC, Kaouk M (1992) Eigenstructure assignment approach for structural damage detection. Am Inst Aeronaut Astronaut J 30:1848–1855

Zio E, Peloni G (2011) Particle filtering prognostic estimation of the remaining useful life of nonlinear components. Reliab Eng Syst Saf 96:403–409

Chapter 2
Data Processing Techniques for Condition Monitoring

2.1 Introduction

As described in Chap. 1, vibration data have been used with varying degrees of success to identify faults in structures (Marwala 2001). Three types of signals have been applied to this end: modal domain *e.g.* the modal properties, frequency domain *e.g.* frequency response functions (FRFs) and time-frequency domain *e.g.* The wavelet transforms (WTs) (Doebling et al. 1996). Marwala and Hunt (1999) applied FRFs and modal properties data simultaneously within the context of a committee of neural networks for fault identification in mechanical structures. Marwala and Heyns (1998) used both modal properties and FRFs in the context of finite element model updating for detecting damage in structures. Marwala (2000) used pseudo-modal energies, modal properties and wavelet data in a committee of neural network for fault identification in a population of cylindrical shells. Marwala (2003) later applied pseudo-modal energies for the classification of faults in structures. Many techniques have been presented that successfully detected faults in structures (Surace and Ruotolo 1994; Manning 1994; Rizos et al. 1990; Stubbs et al. 1992).

In this chapter, modal properties, frequency response functions, pseudo-modal energies, wavelets, mel-frequency cepstral and principal component analysis method are introduced. The next section describes the generalized data acquisition system.

2.2 Data Acquisition System

A generalized data acquisition system is shown in Fig. 2.1. Figure 2.1 demonstrates three main components of the generalized data acquisition system that is implemented in this book for fault identification of faults in a population of cylindrical shells (Marwala 2001):

T. Marwala, *Condition Monitoring Using Computational Intelligence Methods: Applications in Mechanical and Electrical Systems*, DOI 10.1007/978-1-4471-2380-4_2,

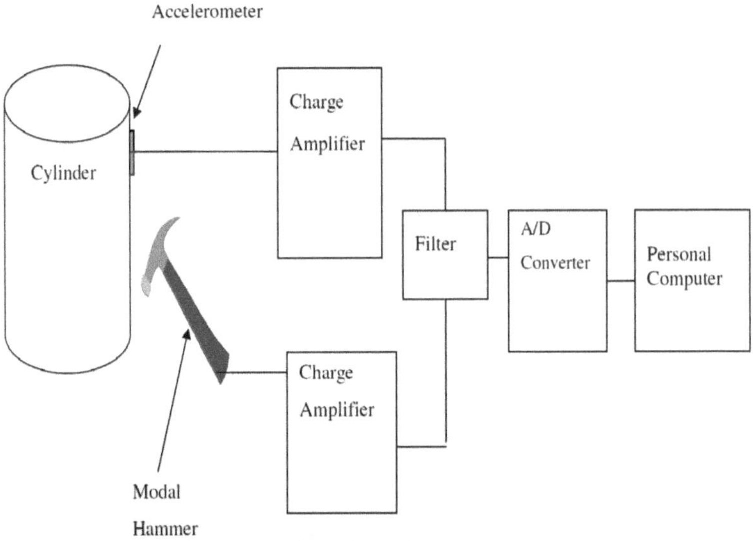

Fig. 2.1 Generalized data acquisition system (Marwala 2001)

1. The excitation mechanism: The aim of the excitation mechanism is to excite the structure so that its response can be measured. In this book for vibration data analysis a modal hammer is applied to excite the structure.
2. The sensing mechanism: The sensing mechanism is used to measure the response from a structure. For example an accelerometer can be applied to measure the acceleration response.
3. The data acquisition and processing: the data is amplified, filtered, converted from analogue to digital format (*i.e.* A/D converter) and finally stored in the computer.

2.2.1 Accelerometers and Impulse Hammer

The cylindrical shells can be excited using a hand-held modal hammer. The modal hammer essentially has three main components: a handle, a force transducer and a hammer tip. The impact of the hammer depends on the mass of the hammer and the velocity of the impact. When such a modal hammer is applied to hit the structure, the operator generally controls the velocity of impact instead of the force itself. The most appropriate technique for fine-tuning the force of the impact is by changing the mass of the hammer. The range of frequencies excited by the hammer depends on the mass of the hammer tip and its stiffness. The hammer tip set-up has a resonance frequency above which it is challenging to deliver energy into the structure and this resonance frequency may be estimated as (contact stiffness / mass of the tip)$^{0.5}$.

For the cylindrical shells examples in this book, the force transducer applied was a PCB A218 and a plastic hammer tip was selected for the reason that it is found to provide sufficient energy to excite frequencies within the bounds of our interest. The sensitivity of the transducer was 4pC/N, and the mass of the head of was 6.6 g. The responses were measured using a DJB piezoelectric accelerometer with a sensitivity of 2.6pC/ms^{-2} and a mass of 19.8 g. A hole of size of 3 mm was drilled into the cylinder and the accelerometer was attached by screwing it through the hole.

2.2.2 *Amplifiers*

Signals from the impulse hammer and the accelerometer provide small charges. As a result the signals needed to be amplified by using a charge amplifier. To achieve this charge amplifiers were designed (Marwala 2001). The acceleration signal was amplified by using a charge amplifier with a sensitivity of 14 mV/pC while the impulse signal was amplified by using a charge amplifier with a sensitivity of 2.0 mV/pC. These amplifiers had a frequency range of 0.44–10 kHz.

2.2.3 *Filter*

A problem related to modal testing is a problem of *aliasing*. When a vibration signal is measured, it must be converted from the analogue into the digital domain as it is sampled by an analogue to digital (A/D) converter. This necessitates that a sampling frequency be selected. If the signal has a substantial difference over a short time then the sampling frequency must be high enough to offer an accurate estimate of a signal that is being sampled. A substantial variation of a signal over a short period of time generally indicates that high frequency components are present in the signal. If the sampling frequency is not high enough, then the high frequency components are not sampled correctly, resulting in the problem called *aliasing,* which is a phenomenon that arises as a result of discretising a signal that was initially continuous. The discretisation process may misinterpret the high frequency components of the signal if the sampling rate is too slow, and this may result in high frequencies looking like low frequencies. During data acquisition, the data were sampled at a rate at least twice the signal frequency to prevent the problem of aliasing. This method is due to the Nyquist-Shannon theorem (Maia and Silva 1997). Additionally, an anti-aliasing filter may be applied before the analogue signal is converted into digital format to circumvent the aliasing problem. An anti-aliasing filter is a low-pass filter which only permits low frequencies to pass through. Fundamentally this filter cuts off frequencies higher than about half of the sampling frequency. For this book, the experiment was performed with a sampling frequency of 10 kHz and the number of samples taken was 8,192. The impulse and the response signals were filtered using a VBF/3 Kemo filter with a gain of 1 and a cut-off frequency of 5 kHz.

2.2.4 Data-Logging System

The National Instruments DAQCard 1,200 with 12-bits over a ±5 V analogue-digital conversion was applied in the cylindrical shell example, to record the impulse force and the acceleration response. The sampling rate was set to 10 kHz, which is adequate for the frequency bandwidth of interest (*i.e.*, 0–5 kHz). A Visual Basic program running on a Daytek desktop computer was used to control the DAQCard. This program was employed to start the data logging, set the sampling frequencies, to check the sample saturation and to save the data. After the raw data were measured and saved, they were then opened using MATLAB and checked as to whether they were acceptable or not by estimating the frequency response functions.

2.3 Fourier Transform

In this section of the book, the Fourier transform was used to calculate the frequency response functions. The Fast Fourier Transform (FFT) is basically a computationally efficient technique for calculating the Fourier transform which exploits the symmetrical nature of the Fourier transform. If the FFT is applied to the response, the following expression is obtained (Ewins 1995):

$$X(\omega) = \frac{1}{2\pi}\int_{-\infty}^{\infty} x(t)e^{-i\omega t}dt \tag{2.1}$$

Similarly, the transformed excitation is (Ewins 1995):

$$F(\omega) = \frac{1}{2\pi}\int_{-\infty}^{\infty} f(t)e^{-i\omega t}dt \tag{2.2}$$

The FRF $\alpha_{ij}(\omega)$ of the response at position i to the excitation at j is the ratio of the Fourier transform of the response to the transform of the excitation (Ewins 1995):

$$\alpha_{ij}(\omega) = \frac{X_i(\omega)}{F_j(\omega)} \tag{2.3}$$

The FRF matrix is related to the spatial properties by the following expression (Ewins 1995):

$$[\alpha(\omega)] = \left[-\omega^2[M] + j\omega[C] + [K]\right]^{-1} \tag{2.4}$$

Here the α is the frequency response function, ω is the frequency, $[M]$ is the mass matrix, $[C]$ is the damping matrix, $[K]$ is the stiffness matrix and $j = \sqrt{-1}$.

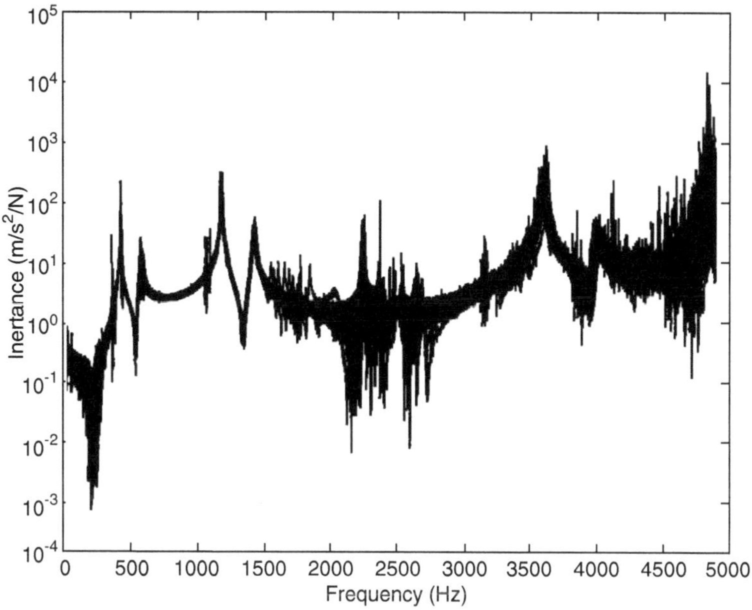

Fig. 2.2 Measured frequency response function of a population of cylinders

Sufficient data to define the relationship between the changes in physical parameters and the changes in FRFs must be generated. From this set of data, a functional mapping between the identity of fault and the FRFs was identified using various computational intelligence tools that are described in this book. An example of a set of the FRFs which were obtained from a population of cylindrical shells is shown in Fig. 2.2 (Marwala 2001).

Sejdic et al. (2011) reviewed the application of the fractional Fourier transform for signal processing and concluded that the main contributions are in digital realizations and its applications. Zhang et al. (2011) applied the fractional Fourier transform to study the propagation properties of Gaussian beams while Harris (1998) applied a Fourier analysis to study biological transients. Nikravan et al. (1989) successfully applied Fourier analysis for the identification of the shapes of simple engineering objects.

2.4 Modal Properties

This section reviews the modal properties which have been applied extensively in fault identification in mechanical systems (Doebling et al. 1996). The modal properties are related to the physical properties of the structure. All elastic structures may be described in terms of their distributed mass, damping and stiffness matrices in the time domain through the following expression (Ewins 1995):

$$[M]\{\ddot{X}\} + [C]\{\dot{X}\} + [K]\{X\} = \{F\} \tag{2.5}$$

Here $\{X\}$, $\{\dot{X}\}$ and $\{\ddot{X}\}$ are the displacement, velocity and acceleration vectors respectively. $\{F\}$ is the applied force. If Eq. 2.5 is transformed into the modal domain to form an eigenvalue equation for the ith mode, then (Ewins 1995):

$$\left(-\bar{\omega}^2[M] + j\bar{\omega}_i[C] + [K]\right)\{\bar{\phi}\}_i = \{0\} \tag{2.6}$$

Here $j = \sqrt{-1}$, $\bar{\omega}_i$ is the ith complex eigenvalue with its imaginary part corresponding to the natural frequency ω_i and is the ith complex mode shape vector with the real part corresponding to the normalized mode shape $\{\phi\}_i$. The sensitivities of the modal properties for undamped case can be written to be (Ewins 1995):

$$\omega_{i,p} = \frac{1}{2\omega_i}\left[\{\phi\}_i^T\left([K]_{,p} - \omega_i^2[M]_{,p}\right)\{\phi\}_i\right] \tag{2.7}$$

and

$$\{\phi\}_{i,p} = \sum_{r=1}^{N}\frac{\{\phi\}_r\{\phi\}_r^T}{\omega_i^2 - \omega_r^2}\left[[K]_{,p} - \omega_i^2[M]_{,p}\right]\{\phi\}_i - \frac{1}{2}\{\phi\}_i\{\phi\}_i^T[M]_{,p}\{\phi\}_i \tag{2.8}$$

In Eqs. 2.7 and 2.8, N is the number of modes, $\omega_{i,p} = \frac{\partial\{\omega\}_i}{\partial g_p}\phi_{i,p} = \frac{\partial\{\phi\}_i}{\partial g_p}[K]_{mp} = \frac{\partial[K]}{\partial g_p}$, $[M]_{mp} = \frac{\partial[M]}{\partial g_p}$ and g_p represents changes in the pth structural parameters. Adhikari (1999) has calculated the damped version of Eqs. 2.7 and 2.8. The introduction of fault in structures changes the mass and stiffness matrices. Equations 2.7 and 2.8 show that changes in the mass and stiffness matrices cause changes in the modal properties of the structure.

Meruane and Heylen (2011) applied modal properties and a hybrid real genetic algorithm to detect structural faults in a tri-dimensional space frame structure. Single and multiple faults scenarios were introduced and the results showed a correct identification of three simultaneous fault locations and gave quantification. Lacarbonara et al. (2007) successfully applied non-linear modal properties to study suspended cables while Lim et al. (2011) successfully applied modal properties to model structural coupling of twin tall buildings with a sky bridge. Eritenel and Parker (2009) applied modal properties to analyze three-dimensional helical planetary gears and found that the modal properties held even for configurations that were not symmetric about the gear plane. Other applications of modal analysis were on solid-core photonic band gap fibers (Viale et al . 2006) and Stamataki et al. (2009) on InGaAsP/InP microring lasers.

2.5 Pseudo-Modal Energies

This book uses pseudo-modal energies for condition monitoring (Marwala 2001). Pseudo-modal energies are the integrals of the real and imaginary components of the frequency response functions over the chosen frequency ranges that bracket the natural frequencies. The frequency response functions may be expressed in receptance and inertance form (Ewins 1995). A *receptance* expression of the frequency response function is defined as the ratio of the Fourier transformed displacement to the Fourier transformed force; while the *inertance* expression of the frequency response function is defined as the ratio of the Fourier transformed acceleration to the Fourier transformed force. This section expresses the pseudo-modal energies in terms of receptance and inertance forms in the same way as the frequency response functions are expressed in these forms.

2.5.1 Receptance and Inertance Pseudo-Modal Energies

The frequency response functions may be expressed in terms of the modal properties by using the modal summation equation (Ewins 1995). Pseudo-modal energies may be estimated as a function of the modal properties from the frequency response functions expressed as a function of modal properties (Marwala 2001). This is performed in order to deduce the capabilities of pseudo-modal energies to identify faults from those of modal properties. The frequency response functions can be expressed in terms of the modal properties using the modal summation equation (Ewins 1995):

$$H_{ki}(\omega) = \sum_{i=1}^{N} \frac{\phi_k^i \phi_l^i}{-\omega^2 + 2j\varsigma_i \omega_i \omega + \omega_i^2} \tag{2.9}$$

Here H_{ki} is the FRF due to excitation at k and measurement at l and ς_i is the damping ratio corresponding to the ith mode. Here it is assumed that the system is proportionally damped. This assumption is valid if the structure being analyzed is lightly damped. Proportional damping is defined as the situation where the viscous damping matrix $[C]$ (see Eq. 2.5) is directly proportional to the stiffness $[K]$ or mass $[M]$ matrix or to the linear combination of both.

The Receptance pseudo Modal Energy (RME) is calculated by integrating the receptance FRF in Eq. 2.9 as follows (Marwala 2001):

$$RME_{kl}^{q} = \int_{a_q}^{b_q} H_{kl} d\omega$$
$$= \int_{a_q}^{b_q} \sum_{i=1}^{N} \frac{\phi_k^i \phi_l^i}{-\omega^2 + 2j\varsigma_i\omega_i\omega + \omega_i^2} d\omega \tag{2.10}$$

In Eq. 2.10, a_q and b_q represent respectively the lower and the upper frequency bounds for the qth pseudo-modal energy. The lower and upper frequency bounds bracket the qth natural frequency. By assuming a light damping ($\varsigma_i \ll 1$), Eq. 2.10 is simplified to give (Gradshteyn and Yyzhik 1994; Marwala 2001)

$$RME_{kl}^{q} \approx \sum_{i=1}^{N} \frac{\phi_k^i \phi_l^i j}{\omega_i} \left\{ \arctan\left(\frac{-\varsigma_i\omega_i - jb_q}{\omega_i} \right) - \arctan\left(\frac{-\varsigma_i\omega_i - ja_q}{\omega_i} \right) \right\} \tag{2.11}$$

The most commonly applied technique to measure vibration data measure the acceleration response instead of the displacement response (Doebling et al. 1996). In such a situation, it is better to calculate the Inertance pseudo-Modal Energies (IMEs) as opposed to the RMEs calculated in Eq. 2.11.

The inertance pseudo-modal energy is derived by integrating the inertance FRF – see (Ewins 1995) for the definition of inertance – expressed in terms of the modal properties by using the modal summation equation (Marwala 2001):

$$IME_{kl}^{q} = \int_{a_q}^{b_q} \sum_{i=1}^{N} \frac{-\omega^2 \phi_k^i \phi_l^i}{-\omega^2 + 2j\varsigma_i\omega_i\omega + \omega_i^2} d\omega \tag{2.12}$$

Assuming that the damping is low, Eq. 2.8 becomes (Gradshteyn and Yyzhik 1994; Marwala 2001):

$$IME_{kl}^{q} \approx \sum_{i=1}^{N} \left[\phi_k^i \phi_l^i \left(b_q - a_q\right) - \omega_i \phi_k^i \phi_l^i j \left\{ \begin{array}{l} \arctan\left(\frac{-\varsigma_i\omega_i - jb_q}{\omega_i} \right) \ldots \\ -\arctan\left(\frac{-\varsigma_i\omega_i - ja_q}{\omega_i} \right) \end{array} \right\} \right] \tag{2.13}$$

Equation 2.13 demonstrates that the inertance pseudo-modal energy may be expressed as a function of the modal properties. The inertance pseudo-modal energies may be estimated directly from the FRFs using any numerical integration scheme. This avoids going through the process of modal extraction.

The advantages of using the pseudo-modal energies over the use of the modal properties are:

- all the modes in the structure are taken into account, as opposed to using the modal properties, which are limited by the number of modes identified; and
- integrating the FRFs to obtain the pseudo-modal energies smooths out the zero-mean noise present in the FRFs.

In this section the pseudo-modal energies were derived mathematically. The next step is to calculate their sensitivities to structural changes with respect to parameter changes.

2.5.2 Sensitivities of Pseudo-Modal Energies

This section assesses the sensitivity of pseudo-modal energies to parameter changes. This offers some insights into how these parameters are affected by the presence of faults. Because the pseudo-modal energies have been derived as functions of the modal properties, these sensitivities are calculated as functions of the sensitivities of the modal properties. The sensitivity of the RMEs are determined by calculating the derivative of Eq. 2.11 with respect to the pth structural changes to give the following expression (Marwala 2001):

$$RME_{kl,p}^{q} \approx \sum_{i=1}^{N} \left\{ \begin{array}{l} \left\{ \frac{j}{\omega_i}\left[\phi_{k,p}^{i}\phi_l^{i} + \phi_k^{i}\phi_{l,p}^{i}\right] - \frac{1}{\omega_i^2} j\phi_k^{i}\phi_l^{i}\omega_{i,p} \right\} \cdots \\ \left\{ \arctan\left(\frac{-\varsigma_i\omega_i - jb_q}{\omega_i}\right) - \arctan\left(\frac{-\varsigma_i\omega_i - ja_q}{\omega_i}\right) \right\} \cdots \\ -\left\{\frac{\phi_k^{i}\phi_l^{i}}{\omega_i}\right\}\left\{\frac{b_q\omega_{i,p}}{\omega_i^2 + \left(\varsigma_i\omega_i + jb_q\right)^2}\right\} - \left\{\frac{a_q\omega_{i,p}}{\omega_i^2 + \left(\varsigma_i\omega_i + ja_q\right)^2}\right\} \end{array} \right\} \tag{2.14}$$

Equation 2.10 is obtained by assuming that ${}^{\partial\varsigma_i}/_{\partial g_p} = 0$ and that $\varsigma_i^2 \approx 0$. For this chapter, faults were introduced by reducing the cross-sectional area of the beam and in later chapters by drilling holes in the structures. Introducing faults this way has been found not to change the damping properties of the structure, thereby justifying the assumption that damping is independent of faults.

Equation 2.14 demonstrates that the sensitivity of the RME is a function of the natural frequencies, the damping ratios, the mode shapes and the derivatives of the natural frequencies and mode shapes. Substituting the derivatives of the modal properties (Adhikari 1999) into Eq. 2.10 gives the sensitivity of the pseudo-modal energies in terms of the mass and stiffness matrices, which are directly related to the physical properties of the structure.

The derivative of the IME with respect to the ith parameter changes may be written as follows (Marwala 2001):

$$IME_{kl,p}^{q} \approx \sum_{i=1}^{N} \left\{ \begin{array}{l} (b_q - a_q)\left(\phi_{k,p}^{i}\phi_l^{i} + \phi_k^{i}\phi_{l,p}^{i}\right) - j\omega_{i,p}\phi_k^{i}\phi_l^{i}\left\{\arctan\left(\dfrac{-\varsigma_i\omega_i - jb_q}{\omega_i}\right)\right\}\ldots \\ -\left\{\arctan\left(\dfrac{-\varsigma_i\omega_i - ja_q}{\omega_i}\right)\right\} - j\omega_i\left(\phi_{k,p}^{i}\phi_l^{i} + \phi_k^{i}\phi_{l,p}^{i}\right)\ldots \\ \left\{\left\{\arctan\left(\dfrac{-\varsigma_i\omega_i}{\omega_i}\right) - \arctan\left(\dfrac{-\varsigma_i\omega_i - ja_q}{\omega_i}\right)\right\}\right\}\ldots \\ +\omega_i\phi_k^{i}\phi_l^{i}\left\{\dfrac{b_q\omega_{i,p}}{\omega_i^2 + (\varsigma_i\omega_i + jb_q)^2} - \dfrac{a_q\omega_{i,p}}{\omega_i^2 + (\varsigma_i\omega_i + ja_q)^2}\right\} \end{array} \right\} \tag{2.15}$$

Similarly, Eq. 2.15 may be expressed in terms of the mass and stiffness matrices by substituting the derivatives of the modal properties (Adhikari 1999) into Eq. 2.15.

In this section the receptance pseudo-modal energies and the inertance pseudo-modal energies were derived and their respective sensitivities were calculated. It was shown how these parameters are related to the modal properties as well as the mass and stiffness matrices. It was found that the sensitivities of the receptance pseudo-modal energies and the inheritance pseudo-modal energies depend upon the sensitivities of the modal properties.

By analyzing the pseudo-modal energies it was observed that if the frequency bandwidth was too narrow, then the energies are dominated by the behavior of the peaks of the FRFs. This is undesirable because near the peaks, factors such as damping ratios, which show high degrees of uncertainty, dominate the dynamics of the pseudo-modal energies. At the same time, if the bandwidth chosen was too wide, the influences of the anti-resonances, which are sensitive to noise, dominate the characteristics of the pseudo-modal energies. An optimal bandwidth is one which is sufficiently narrow to capture the characteristics of the peaks but is wide enough to smooth out the zero-mean noise in the FRFs. It must not be so wide, however, that it includes the anti-resonances.

Equations 2.11–2.15 show that the pseudo-modal energies depend on the modal properties and the frequency bounds chosen. This implies that as long as the FRF information contains the modal properties, then it does not matter how many frequency points are included in the calculation of the pseudo-modal energies. Here it should be noted that the number of frequency points is a separate issue from the frequency bandwidth. On calculating the pseudo-modal energies, the smallest number of frequency points must be applied, and this minimizes the errors in the FRFs that are propagated into the pseudo-modal energies. In other words, for a given frequency bandwidth used in calculating the pseudo-modal energies, increasing the number of frequency points in the bandwidth beyond a certain threshold does not necessarily add any information about the dynamics of the system. It should be noted that the dynamics of the system is the source of information that indicate the presence or the absence of faults.

2.6 Fractal Dimension

Fractal analysis is a technique of describing complex shapes. Many procedures for approximating fractal dimension have been proposed. Lunga and Marwala (2006) successfully applied time-series analysis using fractal theory and online ensemble classifiers to model the stock market while Nelwamondo et al. (2006a) applied fractals for the early classifications of bearing faults. Nelwamondo et al. (2006b) applied a multi-scale fractal dimension for a speaker identification system while (Nelwamondo et al. 2006c) applied fractals for improving speaker identification rates.

Sanchez and Uzcategui (2011) applied fractal geometry in dentistry. They reviewed fractals for treatment and healing monitoring, dental materials, dental tissue, caries, osteoporosis, periodontitis, cancer, Sjogren's syndrome, and the diagnosis of several other conditions. Cross (1994) applied fractal geometric methods for the analysis of microscopic images while Dougherty and Henebry (2001) applied fractal signature and lacunarity to the measurement of the texture of trabecular bone in clinical CT images.

Dathe et al. (2001) applied the fractal dimension method to quantifying soil structures to attain physically based parameters relevant to transport processes while Sokolowska and Sokolowski (2000) applied fractals to study the influence of humic acid on the surface fractal dimension of kaolin.

To define the fractal dimension, let the continuous real-valued function, $s(t), 0 \leq t \leq T$ represents a short-time vibration signal. Furthermore, let the compact planar set represent the graph of this function as follows (Falconer 1952; Nelwamondo et al. 2006b):

$$F = \{(t, s(t) \in R^2 : 0 \leq t \leq T\} \tag{2.16}$$

The fractal dimension of the compact planar set F is called the *Hausdorff dimension* and it is generally between one and two (Falconer 1952). The problem with this dimension is that it is only a mathematical concept and therefore it is tremendously difficult to calculate. So other methods are applied to approximate this dimension such as the Minkowski-Bouligand dimension and the Box-Counting dimension (Falconer 1952). In this book, a fractal dimension is approximated using the Box-Counting dimension, which is discussed in the next section.

2.6.1 Box-Counting Dimension

The Box-Counting dimension (D_B) of, F, is attained by partitioning the plane with a grid of squares each with side ε, and $N(\varepsilon)$ being the number of squares that intersect the plane, defined as (Falconer 1952; Nelwamondo et al. 2006b):

$$D_B(F) = \lim_{\varepsilon \to 0} \frac{\ln N(\varepsilon)}{\ln(1/\varepsilon)} \tag{2.17}$$

Assuming a discrete bearing vibration signal, $s_1, s_2, ..., s_T$ then D_B is given by (Wang et al. 2000; Nelwamondo et al. 2006b):

$$D_B(F) = \left\{ J.\left(\sum_{j=1}^{J} \ln(1/\varepsilon_j).\ln(N(\varepsilon))\right) - \left(\sum_{j=1}^{J} \ln(1/\varepsilon_j)\right).\left(\sum_{j=1}^{J} \ln N(\varepsilon)\right)\right\} / \left\{ J.\sum_{j=1}^{J} (\ln(1/\varepsilon_j))^2 - \left(\sum_{j=1}^{J} \ln(1/\varepsilon)\right)^2 \right\} \tag{2.18}$$

Here J is the computation resolution and $\varepsilon_{\min} \leq \varepsilon_j \leq \varepsilon_{\max}$ with $\varepsilon_{\max}$ and $\varepsilon_{\min}$ represent the maximum and minimum resolutions of computation. In Eq. 2.18, D_B is equal to the slope obtained by fitting a line using least squares method (Maragos and Potamianos 1999).

2.6.2 *Multi-Scale Fractal Dimension (MFD)*

It must be noted that the fractal dimension discussed in the last section is a global measure and consequently does not characterize all the fractal characteristics of the vibration signal (Wang et al. 2000). To deal with this problem of information limitation caused by the global fractal, a Multi-scale Fractal Dimension set is created. The MFD ($D(s,t)$) is obtained by computing the dimensions over a small time window. This MFD set is obtained by dividing the bearing vibration signal into K frames, then K maximum computation resolutions are set as (Wang et al. 2000; Nelwamondo et al. 2006b):

$$\varepsilon_k^{\max} = k.\varepsilon_{\min}(1 \leq k \leq K) \tag{2.19}$$

Here, as before, $\varepsilon_{\min}$ is the minimum valid resolution of the computation. The Box-Counting dimension in Eq. 2.18 can then be written as follows (Falconer 1952; Nelwamondo et al. 2006b):

$$D^k(F) = \left\{ k.\left(\sum_{j=1}^{k} \ln(1/j\varepsilon_{\min}).\ln(N(j\varepsilon_{\min}))\right) - \left(\sum_{j=1}^{k} \ln(1/j\varepsilon_{\min})\right).\left(\sum_{j=1}^{k} \ln N(j\varepsilon_{\min})\right)\right\} / \left\{ k.\sum_{j=1}^{k} (\ln(1/j\varepsilon_{\min}))^2 - \left(\sum_{j=1}^{k} \ln(1/j\varepsilon_{\min})\right)^2 \right\} \tag{2.20}$$

To conclude, the corresponding MFD of the vibration signal is given by (Falconer 1952; Nelwamondo et al. 2006b):

$$MFD(s) = \{D^1(s), D^2(s), ..., D^K(s)\} \tag{2.21}$$

Here $D^k(s)$ is the fractal dimension of the kth frame and this is called the *fractogram* (Wang et al. 2000).

2.7 Mel-Frequency Cepstral Coefficients (MFCCs)

Mel-frequency Cepstral Coefficients (MFCCs) have been widely applied in the field of speech recognition and are can represent the dynamic features of a signal as they extract both linear and non-linear properties. The MFCC can be a useful tool of feature extraction in vibration signals as vibrations contain both linear and non-linear features. The MFCC is a type of wavelet in which frequency scales are placed on a linear scale for frequencies less than 1 kHz and on a log scale for frequencies above 1 kHz (Wang et al. 2002; Nelwamondo et al. 2006a). The complex cepstral coefficients obtained from this scale are called the MFCC (Wang et al. 2002). The MFCC contain both time and frequency information from the signal and this makes them useful for feature extraction. The following types of steps are involved in MFCC computations. Fahmy (2010) applied Mel-frequency Cepstral coefficients for palmprint recognition and their experimental results showed that their technique is robust in the presence of noise. Tufekci et al. (2006) applied Mel-frequency Discrete wavelet coefficients and parallel model compensation for noise-robust speech recognition and the showed improvements of 14.6% and 31.8% error reductions for −6 dB and 0 dB noise levels, respectively.

We transform the input signal, *x(n)* from the time domain to the frequency domain by applying the Fast Fourier Transform (FFT), using (Wang et al. 2002; Nelwamondo et al. 2006a):

$$Y(m) = \frac{1}{F}\sum_{n=0}^{F-1} x(n)w(n)e^{-j\frac{2\pi}{F}nm} \tag{2.22}$$

where F is the number of frames, $0 \leq m \leq F-1$ and $w(n)$ is the Hamming window function given by (Nelwamondo et al. 2006a):

$$w(n) = \beta\left(0.5 - 0.5\cos\frac{2\pi n}{F-1}\right) \tag{2.23}$$

Here $0 \leq n \leq F-1$ and β is the normalization factor defined such that the root mean square of the window is unity (Wang et al. 2002; Nelwamondo et al. 2006a).

Mel-frequency wrapping is performed by changing the frequency to the *mel* using the following equation (Wang et al. 2002; Nelwamondo et al. 2006a):

$$mel = 2595 \times \log_{10}\left(1 + \frac{f_{Hz}}{700}\right) \quad (2.24)$$

Mel-frequency warping uses a filter bank, spaced uniformly on the Mel scale. The filter bank has a triangular band-pass frequency response, whose spacing and magnitude are determined by a constant Mel-frequency interval. The final step converts the logarithmic Mel spectrum back to the time domain. The result of this step is what is called the Mel-frequency Cepstral Coefficients (MFCCs). This conversion is achieved by taking the Discrete Cosine Transform of the spectrum as (Wang et al. 2002; Nelwamondo et al. 2006a):

$$C_m^i = \sum_{n=0}^{F-1} \cos\left(m\frac{\pi}{F}(n + 0.5)\right)\log_{10}(H_n) \quad (2.25)$$

where $0 \leq m \leq L - 1$ and L is the number of MFCC extracted from the ith frame of the signal. H_n is the transfer function of the nth filter on the filter bank. These MFCCs are then applied as a representation of the signal.

2.8 Kurtosis

There is a need to deal with the occasional spiking of vibration data, which is caused by some types of faults. To achieve this task, the method of *kurtosis* is applied. The kurtosis features of vibration data have also been successfully applied in tool condition monitoring by El-Wardany et al. (1996). Kollo (2008) applied multivariate skewness and kurtosis measures in independent component analysis where the solution of an eigenvalue problem of the kurtosis matrix governs the transformation matrix. Furthermore, Antoni (2006) applied kurtosis for the characterising of non-stationary signals while de la Rosa et al. (2010) applied kurtosis for the non-destructive measurement of termite activity and Boumahdi (1996) applied kurtosis for the blind identification of a field seismic data. The success of kurtosis in vibration signals is based on the fact that a system which is under stress or has defects has vibration signals which differ from those of a normal system. The sharpness or spiking of the vibration signal changes when there are defects in the system. Kurtosis is a measure of the sharpness of the peak and is defined as the normalized fourth-order central moment of the signal (Wang et al. 2001). The kurtosis value is useful in identifying transients and spontaneous events within vibration signals (Wang et al. 2001) and is one of the accepted criteria in fault detection. The calculated kurtosis value is typically normalized by the square of the second moment. A high value of kurtosis implies a sharp distribution peak and indicates that the signal is impulsive in nature (Altman and Mathew 2001).

$$K = \frac{1}{N}\sum_{i=1}^{N}\frac{(x_i - \bar{x})^4}{\sigma^4} \tag{2.26}$$

where $\bar{x}$ is the mean and σ is the variance.

2.9 Wavelet Transform

Cheng et al. (2011) applied wavelets for the spectroscopic determination of leaf water content while Magosso et al. (2009) applied a wavelet-based energetic approach to an electroencephalogram and an electro-oculogram. Other successful applications of wavelet data include fault detection in frame structures (Ovanesova and Suarez 2004), hypothesis testing of brain activation maps (Fadili and Bullmore 2004), echocardiographic texture analysis (Kerut et al. 2000), to describe embolic signals (Aydin et al. 1999) as well as for fault detection of coherent structures (Gilliam et al. 2000).

The Wavelet Transform (WT) of a signal is an illustration of a timescale decomposition which highlights the local features of a signal. Wavelets occur in sets of functions that are defined by *dilation*, which controls the scaling parameter, and *translation*, which controls the position of a single function known as the *mother wavelet* $w(t)$. In general, each set of wavelets can be written as follows (Marwala 2000; Newland 1993):

$$W_{ab}(t) = \frac{1}{\sqrt{a}} w\left(\frac{t-b}{a}\right) \tag{2.27}$$

Here b = translation parameter, which localizes the wavelet function in the time domain; a = dilation parameter, defining the analyzing window stretching; and w = mother wavelet function. The continuous WT of a signal $x(t)$ is defined as (Marwala 2000; Newland 1993):

$$W\left(2^j + k\right) = 2^j \int_{-\infty}^{\infty} x(t) w^*(2^j t - k) dt \tag{2.28}$$

Here w^* = complex conjugate of the basic wavelet function; j is called the *level* (scale), which determines how many wavelets are needed to cover the mother wavelet and is the same as a frequency varying in harmonics and k determines the position of the wavelet and gives an indication of time. The length of the data in the time domain must be an integer power of two. The wavelets are organized into a sequence of levels 2^j, where j is from 1 to n-1. Equations 2.27 and 2.28 are valid for

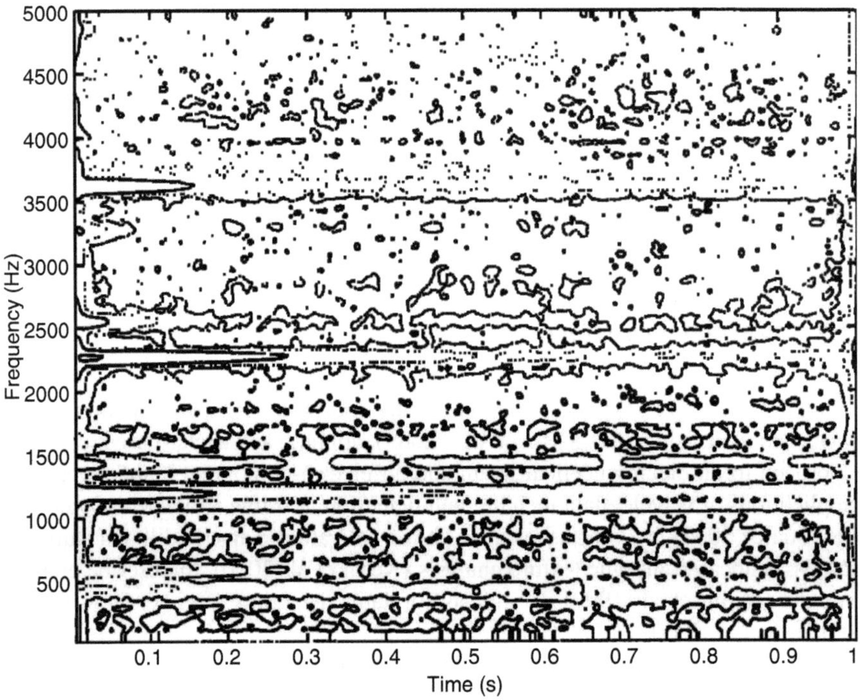

Fig. 2.3 Wavelet spectrum from one cylinder

$0 \leq k$ and $0 \leq k \leq 2^j$-1. The WT in this book is from the orthogonal wavelet family (Daubechie 1991) defined by Newland (1993) as follows (Marwala 2000):

$$w(t) = \frac{\left(e^{i4\pi t} - e^{i2\pi t}\right)}{i2\pi t} \tag{2.29}$$

The WT may also be formulated by transforming the signal $x(t)$ and the wavelet function into the frequency domain as follows (Marwala 2000):

$$W\,(j,k) = \int_{2\pi 2^j}^{4\pi 2^j} X(\omega) e^{i\omega k l 2^j} d\omega \tag{2.30}$$

The relationship between the physical properties of the structure and the WT of the impulse of unit magnitude may be applied to identify faults in structures. Liew and Wang (1998) applied such WT data to identify faults in structures. A functional mapping between the identity of a fault and the WT of the response k may be identified. The transform data for a wavelet is shown in Fig. 2.3 (Marwala 2000).

2.10 Principal Component Analysis

For this book the principal component analysis (PCA) (Jolliffe 1986) was implemented to reduce the input data into independent components. The PCA orthogonalizes the components of the input vector so that they are uncorrelated with each other. When implementing the PCA for data reduction, correlations and interactions among variables in the data are summarised in terms of a small number of underlying factors. The PCA was introduced by Pearson (1901) to recast linear regression analysis into a new framework, and was developed further by Hotelling (1933) who applied it to Psychometry and it was subsequently generalized by Loéve (1963). The PCA has been successfully applied to reduce the dimension of the data (Bishop 1995). Some researchers who have successfully applied this technique include Partridge and Calvo (1997) who applied the PCA to reduce the dimensions of two high-dimensional image databases, one of handwritten digits and one of handwritten Japanese characters. The variant of the PCA used in this book finds the directions in which the data points have the most variance. These directions are called *principal directions*. The data are then projected onto these principal directions without the loss of significant information from the data. Here follows a brief outline of the implementation of the PCA adopted in this book. The covariance matrix can be calculated as follows (Marwala 2001):

$$\Sigma = \sum_{p=1}^{P} (x^p - \mu)(x^p - \mu)^T \tag{2.31}$$

Here Σ is the covariance matrix, T is for transpose, P is the number of vectors in the training set, μ is the mean vector of the data set taken over the number of training set and x is the input data. The second step is to calculate the eigenvalues and eigenvectors of the covariance matrix and arrange them from the largest eigenvalue to the smallest. The first N largest eigenvalues are chosen. In this book the first N eigenvalues were chosen in such a way that their sum constitutes at least 85% of the total sum of all the eigenvalues. By so doing at least 85% of the variance of the data was retained. The data were then projected onto the eigenvectors corresponding to the N most dominant eigenvalues.

2.11 Examples Used in This Book

This section reports on three examples: a rolling element bearing, a population of cylindrical shells and transformer bushings.

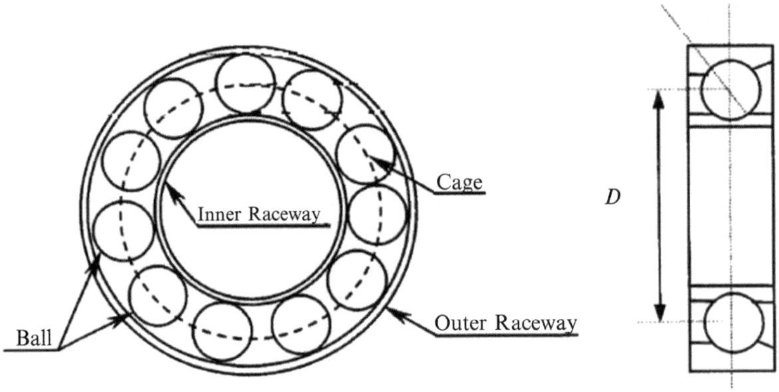

Fig. 2.4 Ball bearing geometry, with diameter D

2.11.1 Rolling Element Bearing

Vibration measurement is important in an advanced conditioning monitoring of mechanical systems. Most bearing vibrations are periodical movements. The geometry of the bearing is shown in Fig. 2.4 (Purushothama et al. 2005; Li et al. 2000; Nelwamondo et al. 2006a; Marwala et al. 2006). Generally, the rolling bearing contains two concentric rings, which are called the *inner* and *outer raceway* (Li et al. 2000). Furthermore, the bearing contains a set of rolling elements that run in the tracks of these raceways. There are number of standard shapes for the rolling elements such as the ball, the cylindrical roller, the tapered roller, needle rollers, the symmetrical and the unsymmetrical barrel roller (Li et al. 2000). For this book, a ball rolling element was used. Figure 2.4 also shows the cage, which ensures uniform spacing and prevents mutual contact.

There were three faults studied for this book: an inner raceway fault, an outer raceway fault and a rolling element fault. A bearing fault increases the rotational friction of the rotor and therefore each fault generates vibration spectra with unique frequency components (Ericsson et al. 2004). It should be noted that these frequency components are a linear function of the running speed. Additionally, the two raceway frequencies are also linear functions of the number of balls. The motor-bearing conditioning monitoring systems were implemented by analyzing the vibration signal of all the bearing faults. The vibration signal was produced by the impact pulse generated when a ball roller hit a defect in the raceways or each and every time the defect in the ball hit the raceways (Li et al. 2000).

The database used to validate new bearing fault diagnosis was developed at the Rockwell Science Center by Loparo (Purushothama et al. 2005; Lou and Loparo 2004) In this database, single point faults of diameters of 7 mils, 14 mils and 21 mils (1 mil = 0.001 in.) were introduced using electro-discharge machining. These faults were introduced separately at the inner raceway, rolling element (*i.e.*, ball) and outer raceway. The experiments were performed for each fault diameter and

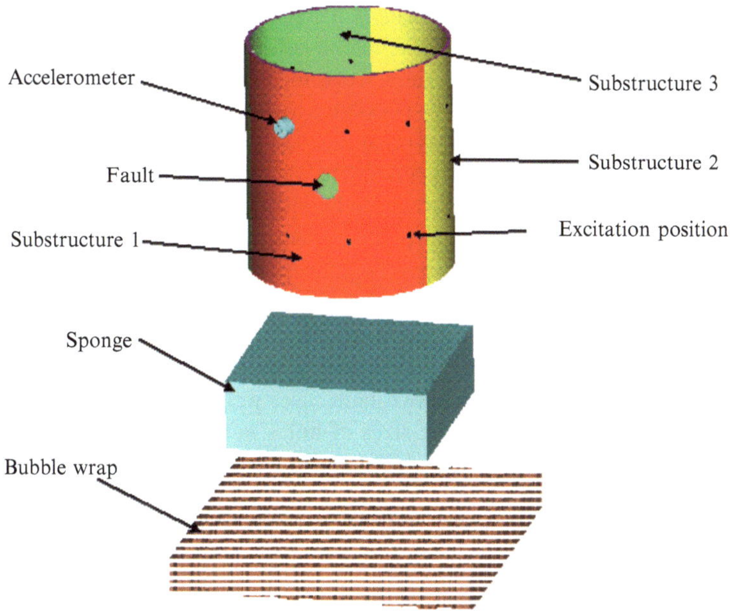

Fig. 2.5 Illustration of the cylindrical shell showing the positions of the excitation impulse, accelerometer, substructures, fault position and supporting sponge

this was repeated for two load conditions, which were 1 and 2 horsepower. The experimentation was performed for vibration signals sampled at 12,000 samples per second for the drive-end bearing faults. The vibration signals from this database were all divided into equal windows of four revolutions. Half of the resulting sub-signals were used for training and the other half were used for testing.

2.11.2 Population of Cylindrical Shells

The second experiment was performed on a population of cylinders, which were supported by inserting a sponge rested on a bubble-wrap, to simulate a 'free-free' environment (see Fig. 2.5). The impulse hammer test was performed on each of the 20 steel seam-welded cylindrical shells. The impulse was applied at 19 different locations as indicated in Fig. 2.5. More details on this experiment may be found in Marwala (2001).

Each cylinder was divided into three equal substructures and holes of 10–15 mm diameter were introduced at the centers of the substructures to simulate faults. For one cylinder the first type of fault was a zero-fault scenario. This type of fault was given the identity [000], indicating that there were no faults in any of the three substructures. The second type of fault was a one-fault-scenario, where a hole could be located in any of the three substructures. Three possible one-fault-scenarios were [100], [010], and [001] indicating one hole in substructures 1, 2 or 3, respectively. The third type of fault was a two-fault scenario, where holes were located in two of the three substructures. The three possible two-fault-scenarios were identified as [110], [101], and [011]. The final type of fault was a three-fault-scenario, where a hole was located in all three substructures. The identity of this fault was [111]. Eight types of fault-cases were considered (including [000]).

Each cylinder was measured three times under different boundary conditions by changing the orientation of the rectangular sponge inserted inside the cylinder. The number of sets of measurements taken for undamaged population was 60 (20 cylinders × 3 different boundary conditions). For the eight possible fault types, two fault types [000] and [111] had 60 occurrences while the rest had 24. It should be noted that the numbers of one- and two-fault cases were each 72. This was because as mentioned above, increasing the sizes of holes in the substructures and taking vibration measurements generated additional one- and two-fault cases.

2.11.3 Transformer Bushings

Bushings are an important component in electricity for transportation. They are used in substation buildings, transformers, locomotives, and switchgear. Bushings cause more than 15% of transformer failures in Eskom (van Wyk 1997). Australasian reliability statistics on transformers over 1970–1995 concluded that bushings were second to tap changers as the component initially involved in failure and were amongst the top three contributors to costly transformer failures (Lord and Hodge 2003). Figure 2.6 shows that the result of collateral damage and personnel injury is a major concern, warranting the development of new and improved, affordable, reliable, and flexible diagnostics technologies to allow asset owners to detect impending failures in a timely manner (Dhlamini and Marwala 2004a). Sokolov (2001) found that more than 46% of transformer defects were attributable to bushings, on-load tap changers, and the cooling system. The reliability of bushings affects the security of supply of electricity in an area and the economical operation of the area. Transformer failure research shows that bushings are among the top three most frequent causes of transformer failure (Ward 2000; Lord and Hodge 2003; Vilakazi et al. 2005). Bushing failure is typically followed by a catastrophic event such as a tank rupture, violent explosion of the bushing and fire (Lord and Hodge 2003). With such consequences the major concern is collateral faults and personnel injury. Numerous diagnostic tools exist such as on-line Partial Discharge analysis, on-line power factor, and infra-red scanning to detect an impending transformer failure (Mojo 1997). In isolation, few of these approaches can offer all of the data that a transformer operator needs to decide upon a course of action. Condition monitoring has many benefits such as: an unexpected failure can be avoided through the possession of quality information relating to on-line condition of the plant and the resulting ability to identify faults in incipient levels of development.

Fig. 2.6 Limited fault due to exploded bushing (Dhlamini 2007)

Computational intelligence methods can be used for bushing condition monitoring. For this book, approaches based on computational intelligence techniques were developed and then used for interpreting the data from a Dissolve Gas-in-oil Analysis (DGA) test. A DGA is the most commonly used diagnostic technique for transformers and bushings (Dhlamini and Marwala 2004b; Ding et al. 2000). A DGA is used to detect oil breakdown, moisture presence and PD activity. Fault gases are produced by the degradation of the transformer and bushing oil and solid insulation such as paper and pressboard, which are all made of cellulose (Saha 2003). The gases produced from the transformer and bushing operation are (Yanming and Zheng 2000; Castro and Miranda 2004; Dhlamini and Marwala 2004b; Vilakazi et al. 2005):

- Hydrocarbons gases and hydrogen: methane (CH_4), ethane (C_2H_6), ethylene (C_2H_4), acetylene (C_2H_2) and hydrogen (H_2);
- Carbon oxides: carbon monoxide (CO) and carbon dioxide (CO_2); and
- Non-fault gases: nitrogen (N_2) and oxygen (O_2).

The causes of faults fall into two main groups: partial discharges and thermal heating faults. Partial discharge faults are further divided into high-energy discharge and low-energy discharge faults. High-energy discharge is known as *arcing* and low energy discharge is referred to as *corona*. The quantity and types of gases reflect the nature and extent of the stressed mechanism in the bushing (Zhang 1996). Oil breakdown is shown by the presence of hydrogen, methane, ethane, ethylene and acetylene. High levels of hydrogen demonstrate that the degeneration is due to corona. High levels of acetylene occur in the presence of arcing at high temperature. Methane and ethane are produced from low- temperature thermal heating of oil but high-temperature thermal heating produces ethylene and hydrogen plus methane and ethane. The low-temperature thermal degradation of cellulose produces CO_2 and high temperature heating produces CO.

2.12 Conclusions

In this chapter, the data processing techniques for condition monitoring in mechanical and electrical systems were reviewed. Approaches for acquiring data were described and procedures for analyzing data were explained. Modal properties, pseudo-modal energies, wavelet, principal component analysis and Mel-frequency Cepstral Coefficients methods were described. Cases used in this book were described: gearbox data, population of cylindrical shells data and transformer bushing data.

References

Adhikari S (1999) Rates of change of eigenvalues and eigenvectors in damped system. Am Inst Aeronaut Astronaut J 37:1452–1458

Altman J, Mathew J (2001) Multiple band-pass autoregressive demodulation for rolling element bearing fault diagnosis. Mech Syst Signal Process 15(5):963–997

Antoni J (2006) The spectral kurtosis: a useful tool for characterising non-stationary signals. Mech Syst Signal Process 20(2):282–307

Aydin N, Padayachee S, Markus HS (1999) The use of the wavelet transform to describe embolic signals. Ultrasound Med Biol 25(6):953–958

Bishop CM (1995) Neural networks for pattern recognition. Oxford University Press, Oxford

Boumahdi M (1996) Blind identification using the kurtosis with applications to field data. Signal Process 48(3):205–216

Castro ARG, Miranda V (2004) An interpretation of neural networks as inference engines with application to transformer failure diagnosis. In: IEEE international conference on probabilistic methods applied to power systems, Porto, Portugal, pp 997–1002

Cheng T, Rivard B, Sanchez-Azofeifa A (2011) Spectroscopic determination of leaf water content using continuous wavelet analysis. Remote Sens Environ 115(2):659–670

Cross SS (1994) The application of fractal geometric analysis to microscopic images. Micron 25(1):101–113

Dathe A, Eins S, Niemeyer J, Gerold G (2001) The surface fractal dimension of the soil-pore interface as measured by image analysis. Geoderma 103(1–2):203–229

Daubechie I (1991) The wavelet transform, time-frequency localization, and signal processing. IEEE Trans Inform Theory 36:961–1005

de la Rosa JJG, Moreno-Munoz A, Gallego A, Piotrkowski R, Castro E (2010) On-site non-destructive measurement of termite activity using the spectral kurtosis and the discrete wavelet transform. Measurement 43(10):1472–1488

Dhlamini SM (2007) Bushing diagnosis using artificial intelligence and dissolved gas analysis. PhD thesis, University of the Witwatersrand

Dhlamini SM, Marwala T (2004a) An application of SVM, RBF and MLP with ARD on bushings. In: Proceedings of the IEEE conference on cybernetics and intelligent systems, Singapore, pp 1245–1258

Dhlamini SM, Marwala T (2004b) Bushing monitoring using MLP and RBF. In: Proceedings of the IEEE Africon 2004, Gaborone, Botswana, pp 613–617
Ding X, Liu Y, Griffin PJ (2000) Neural nets and experts system diagnoses transformer faults. IEEE Comput Appl Power 13(1):50–55
Doebling SW, Farrar CR, Prime MB, Shevitz DW (1996) Damage identification and health monitoring of structural and mechanical systems from changes in their vibration characteristics: a literature review Los Alamos National Laboratory report LA-13070-MS
Dougherty G, Henebry GM (2001) Fractal signature and lacunarity in the measurement of the texture of trabecular bone in clinical CT images. Med Eng Phys 23(6):369–380
El-Wardany TI, Gao D, Elbestawi MA (1996) Tool condition monitoring in drilling using vibration signature analysis. Int J Mach Tool Manufacture 36(6):687–711
Ericsson S, Grip N, Johansson E, Persson LE, Sjoberg R, Stromberg JO (2004) Towards automatic detection of local bearing defects in rotating machines. Mech Syst Signal Process 19:509–535
Eritenel T, Parker RG (2009) Modal properties of three-dimensional helical planetary gears. J Sound Vib 325(1–2):397–420
Ewins DJ (1995) Modal testing: theory and practice. Research Studies Press, Letchworth
Fadili MJ, Bullmore ET (2004) A comparative evaluation of wavelet-based methods for hypothesis testing of brain activation maps. NeuroImage 23(3):1112–1128
Fahmy MMM (2010) Palmprint recognition based on Mel frequency Cepstral coefficients feature extraction. Ain Shams Eng J 1(1):39–47
Falconer K (1952) Fractal geometry; mathematical foundations and application. John Wiley, New York
Gilliam X, Dunyak J, Doggett A, Smith D (2000) Coherent structure detection using wavelet analysis in long time-series. J Wind Eng Ind Aerodyn 88(2–3):183–195
Gradshteyn IS, Yyzhik IM (1994) Tables of integrals, series, and products. Academic, London
Harris CM (1998) The Fourier analysis of biological transients. J Neurosci Meth 83(1):15–34
Hotelling H (1933) Analysis of a complex of statistical variables into principal components. J Educ Psychol 24:417–441
Jolliffe IT (1986) Principal component analysis. Springer, New York
Kerut EK, Given MB, McIlwain E, Allen G, Espinoza C, Giles TD (2000) Echocardiographic texture analysis using the wavelet transform: differentiation of early heart muscle disease. Ultrasound Med Biol 26(9):1445–1453
Kollo T (2008) Multivariate skewness and kurtosis measures with an application in ICA. J Multivar Anal 99(10):2328–2338
Lacarbonara W, Paolone A, Vestroni F (2007) Non-linear modal properties of non-shallow cables. Int J Non-Linear Mech 42(3):542–554
Li B, Chow MY, Tipsuwan Y, Hung JC (2000) Neural-network-based motor rolling bearing fault diagnosis. IEEE Trans Ind Electron 47:1060–1068
Liew KM, Wang Q (1998) Application of wavelet theory for crack identification in structures. J Eng Mech ASCE 124(2):152–157
Lim J, Bienkiewicz B, Richards E (2011) Modeling of structural coupling for assessment of modal properties of twin tall buildings with a skybridge. J Wind Eng Ind Aerodyn 99(5):615–623
Loéve M (1963) Probability theory, 3rd edn. Van Nostrand, New York
Lord T, Hodge G (2003) On-line monitoring technology applied to HV bushing. In: Proceedings of the AVO conference, New Zealand, November (CD-ROM)
Lou X, Loparo KA (2004) Bearing fault diagnosis based on wavelet transform and fuzzy inference. J Mech Syst Signal Process 18:1077–1095
Lunga D, Marwala T (2006) Time series analysis using fractal theory and online ensemble classifiers, vol 4304/2006, Lectures notes in artificial intelligence. Springer, Berlin/Heidelberg, pp 312–321
Magosso E, Ursino M, Zaniboni A, Gardella E (1 January 2009) A wavelet-based energetic approach for the analysis of biomedical signals: application to the electroencephalogram and electro-oculogram. Appl Math Comput, Includes Special issue on Emergent Applications of Fractals and Wavelets in Biology and Biomedicine 207(1):42–62

Maia NMM, Silva JMM (1997) Theoretical and experimental modal analysis. Research Studies Press, Letchworth
Manning R (1994) Damage detection in adaptive structures using neural networks. In: Proceedings of the 35th AIAA/ASME/ASCE/AHS/ASC structures, structural dynamics, and materials conference, Hilton Head, South Carolina, pp 160–172
Maragos P, Potamianos A (1999) Fractal dimensions of speech sounds: computation and application to automatic speech recognition. J Acoust Soc Am 105:1925–1932
Marwala T (2000) On damage identification using a committee of neural networks. American Society of Civil Engineers. J Eng Mech 126:43–50
Marwala T (2001) Fault identification using neural networks and vibration data. Doctor of Philosophy, University of Cambridge
Marwala T (2003) Fault classification using pseudo modal energies and neural networks. Am Inst Aeronaut Astronaut J 41(1):82–89
Marwala T, Heyns PS (1998) A multiple criterion method for detecting damage on structures. Am Inst Aeronaut Astronaut J 195:1494–1501
Marwala T, Hunt HEM (1999) Fault identification using a committee of neural networks. In: Friswell MI, Mottershead JE, Lees AW (eds) Identification in engineering systems. Swansea Wales, UK, pp 102–111
Marwala T, Mahola U, Nelwamondo F (2006) Hidden Markov models and Gaussian mixture models for bearing fault detection using fractals. In: Proceedings of the IEEE international joint conference on neural networks, Vancouver, BC, Canada, pp 5876–5881
Meruane V, Heylen W (2011) An hybrid real genetic algorithm to detect structural damage using modal properties. Mech Syst Signal Process 25(5):1559–1573
Mojo B (1997) Transformer condition monitoring (non-invasive Infrared Thermography Technique). Masters thesis, Department of Electrical Engineering, University of the Witwatersrand
Nelwamondo FV, Marwala T, Mahola U (2006a) Early classifications of bearing faults using hidden Markov models, Gaussian mixture models, Mel-frequency Cepstral coefficients and fractals. Int J Innov Comput Inform Control 2(6):1281–1299
Nelwamondo FV, Mahola U, Marwala T (2006b) Multi-scale fractal dimension for speaker identification system. Trans Syst 5(5):1152–1157
Nelwamondo FV, Mahola U, Marwala T (2006c) Improving speaker identification rate using fractals. In: Proceedings of the IEEE international joint conference on neural networks, Vancouver, BC, Canada, pp 5870–5875
Newland DE (1993) An introduction to random vibration, spectral and wavelet analysis, 3rd edn. Longman, Harlow, and John Wiley, New York
Nikravan B, Baul RM, Gill KF (1989) An experimental evaluation of normalised Fourier descriptors in the identification of simple engineering objects. Comput Ind 13(1):37–47
Ovanesova AV, Suarez LE (2004) Applications of wavelet transforms to damage detection in frame structures. Eng Struct 26(1):39–49
Partridge MG, Calvo RA (1997) Fast dimensionality reduction and simple PCA. Intell Data Anal 2(3):724–727
Pearson K (1901) On lines and planes of closest fit to systems of points in space. Philos Mag 2:559–572
Purushothama V, Narayanana S, Prasadb SAN (2005) Multi-fault diagnosis of rolling bearing elements using wavelet analysis and hidden Markov model based fault recognition. NDT&E Int 38:654–664
Rizos PF, Aspragathos N, Dimarogonas AD (1990) Identification of crack location and magnitude in a cantilever beam from the vibration modes. J Sound Vib 138:381–388
Saha TK (2003) Review of modern diagnostic techniques for assessing insulation condition in aged transformer. IEEE Trans Dielect Electr Insul 10(5):903–917
Sanchez I, Uzcategui G (2011) Fractals in dentistry. J Dent 39(4):273–292
Sejdic E, Djurovic I, Stankovic L (2011) Fractional fourier transform as a signal processing tool: an overview of recent developments, signal processing. Fourier Related Transforms for Non-Stationary Signals 91(6):1351–1369

Sokolov V (2001) Transformer life management, II workshop on power transformers-deregulation and transformers technical, economic, and strategical issues, Salvador, Brazil, 29–31 August
Sokolowska Z, Sokolowski S (2000) Influence of humic acid on surface fractal dimension of kaolin: analysis of mercury porosimetry and water vapour adsorption data. In: Pachepsky Y, Crawford JW, Rawls WJ (eds) Fractals in soil science, vol 27, Developments in soil science. Elsevier, Amsterdam/New York, pp 143–159
Stamataki I, Kapsalis A, Mikroulis S, Syvridis D, Hamacher M, Troppenz U, Heidrich H (2009) Modal properties of all-active InGaAsP/InP microring lasers. Opt Commun 282(12):2388–2393
Stubbs NJ, Kim JT, Topole K (1992) An efficient and robust algorithm for damage localization in offshore platforms. In: Proceeding of the American Society of Civil Engineers 10th structures congress, San Antonio, pp 543–546
Surace C, Ruotolo R (1994) Crack detection of a beam using the wavelet transform. In: Proceedings of the 12th international modal analysis conference, Honolulu, Hawaii, pp 1141–1148
Tufekci Z, Gowdy JN, Gurbuz S, Patterson E (2006) Applied mel-frequency discrete wavelet coefficients and parallel model compensation for noise-robust speech recognition. Speech Commun 48(10):1294–1307
van Wyk S (1997) Transformer field performance in subSaharan Africa. Energize Magazine, Jul/Aug issue:44–47
Viale P, Fevrier S, Leproux P, Jaouen Y, Obaton A-F (2006) Modal properties of solid-core photonic bandgap fibers. Photonics Nanostruct – Fundam Appl 4(2):116–122
Vilakazi CB, Mautla RP, Moloto EM, Marwala T (2005) On-line bushing condition monitoring. In: Garcia-Planas MI et al (eds) Proceedings of the 5th WSEAS/IASME international conference on electric power systems, high voltages, electric machines, Tenerife, Spain, pp 406–411
Wang F, Zheng F, Wu W (2000) A C/V segmentation for Mandarin speech based on multi-scale fractal dimension. In: International conference on spoken language processing, Beijing, China, pp 4:648–651
Wang J, Wang J, Weng Y (2002) Chip design of MFCC extraction for speech recognition. Integr VLSI J 32(1–3):111–131
Wang Z, Willett P, DeAguiar PR, Webster Y (2001) Neural network detection of grinding burn from acoustic emission. Int J Mach Tool Manufacture 41(2):283–309
Ward B (2000) A survey of new techniques in insulation monitoring of power transformers. IEEE Electr Insul Mag 17(3):16–23
Yanming T, Zheng Q (2000) DGA based insulation diagnosis of power transformer via ANN. In: Proceedings of the 6th international conference on properties and application of dielectric materials, Winnipeg, Canada, pp 1:133–137
Zhang J, Xu Q, Lu X (2011) Propagation properties of Gaussian beams through the anamorphic fractional Fourier transform system with an eccentric circular aperture. Optik – Int J Light Electron Opt 122(4):277–280
Zhang Y (1996) An artificial neural network approach to transformer fault diagnosis. IEEE Trans Power Deliv 11:1836–1841

Chapter 3
Multi-Layer Perceptron for Condition Monitoring in a Mechanical System

3.1 Introduction

It is important to identify faults before the occurrence of faults in a mechanical system as it prevents the loss of life and reduces the machine down-time. The process of fault identification entails the gathering of data, the processing of data to reveal vital features and the interpretation of the data. For this chapter, vibration data were measured and were processed as explained in Chap. 2. The data was then analyzed to identify the faults. The mechanism that was adopted to analyze the data in this chapter was the Multi-Layer Perceptron (MLP) neural network.

Dimla and Lister (2000) applied the MLP neural network for the monitoring of the condition of an on-line metal cutting tool. They performed test cuts on an EN24 alloy steel using P15 and P25 coated cemented carbide inserts. Thereafter, cutting forces and vibration data were measured online. Concurrently the wear sizes on the cutting boundaries were measured and these as well as the processed data were put into a multi-layer perceptron neural network that had been trained to identify the tool state. When the proposed system was tested on a number of cutting types it was able to classify tool state condition with an accuracy of 90%. However, the performance worsened when the cutting types were altered considerably.

Mustapha et al. (2007) applied a vector of novelty indices for damage detection and location in an isotropic plate. Their method was based on the novelty detection technique, outlier analysis and a multi-layer perceptron neural network. To evaluate the usefulness of the method, a thin rectangular plate with isotropic behavior was experimentally assessed. Additionally, a study was made of the scattering effect of an ultrasonic guided wave on the investigated plate, for both a faulty and a healthy state of affairs. Wave propagation was successively communicated and measured by 8 Piezo-electic Transducer (PzT) patches bonded on the plate, creating a sensor network on the tested isotropic rectangular structure. An 8-channel multiplexer was included in the small scale, low-cost 'structural health monitoring' system to switch the PzTs task from sensor to actuator. Fault identification software was developed to obtain the waveform responses. The scattering waveform responses indicating

T. Marwala, *Condition Monitoring Using Computational Intelligence Methods: Applications in Mechanical and Electrical Systems*, DOI 10.1007/978-1-4471-2380-4_3,

healthy and faulty conditions were transformed into a set of novelty indices that eventually define the conditions of the tested structure. The developed novelty indices indicating the available sensor paths were applied as the inputs to the MLP neural network to identify faults on the tested isotopic plate.

Herzog et al. (2009) applied the MLP neural network for estimating the residual life of a machine and its components. They trained and tested numerous neural network variations with data from two different reliability-related datasets. The first dataset represented the renewal case where the failed unit was repaired and restored to a good-as-new condition. Data were collected in the laboratory by subjecting a series of similar test pieces to fatigue loading with a hydraulic actuator. The average prediction error of the various neural networks being compared varied from 431 to 841 s on this dataset, where test pieces had a characteristic life of 8971 s. The second dataset was gathered from a collection of pumps used to circulate a water and magnetite solution within a plant. The data was obtained from a repaired system affected by reliability degradation. When optimized, the multi-layer perceptron neural networks trained with the Levenberg-Marquardt procedure and the general regression neural network gave a sums-of squares error within 11.1% of each other for the renewal dataset. The small number of inputs and poorly mapped input space on the second dataset indicated that much larger errors were verified on some of the test data. Nevertheless, the prospect of using neural networks for residual life prediction and the advantage of integrating condition-based data into the model was demonstrated in both examples.

Rafiee et al. (2007) applied a multi-layer perceptron neural network for the intelligent condition monitoring of a gearbox. The input to the MLP neural network consisted of the standard deviations of wavelet packet coefficients of vibration data. The gear conditions were classified as normal gearbox, slightly-worn, medium-worn, broken-teeth gear damage and a general bearing fault. The results indicated an accurate identification of the faults.

Abu-Mahfouz (2003) successfully applied vibration data and multi-layer perceptron neural network for detection and classification of drilling wear. The study compared several architectures and used a vibration signature as the source of information from the machining process. Five different drill wear conditions were simulated and used to train the neural network for fault identification. It was observed that the frequency domain features were more effective in training the neural network than were the time domain data.

Kim et al. (2006) applied wavelets with Daubechie's four functions and multi-layer perceptron network neural networks to detect the toxic response behavior of chironomids for water quality monitoring. The variables chosen were based on the feature coefficients of Discrete Wavelet Transforms and were used as input for training the MLP network. The trained network capably detected changes in movement patterns before and after the treatments and it was shown that the application of the wavelets and artificial neural networks was useful for the automatic monitoring of water quality.

Marwala and Hunt (1999) presented a committee of neural networks technique, which employed both frequency response functions and modal data simultaneously to identify faults in structures. They tested this technique on simulated data from a cantilevered beam, which was structured into five regions. It was observed that irrespective of the noise levels in the data, the committee of neural networks gave results that had a lower mean-squares error and standard deviation than the two existing methods. It was found that the method could identify fault cases better than the two approaches used individually. It was established that for the problem analyzed, giving equal weights to the frequency-response-based method and modal-properties-based method minimized the errors in identifying faults.

Bucolo et al. (2002) successfully applied multi-layer perceptron and neuro-fuzzy systems in predicting models for the corrosion phenomena in pulp and paper plant. The data was gathered in a Wisaforest pulp mill in Finland.

Caputo and Pelagagge (2002) successfully applied the MLP neural network approach for piping network monitoring to localize leakages based on pressure and flow rate information. Elsewhere Yella et al. (2009) successfully applied machine vision based method for the condition monitoring of wooden railway sleepers. Furthermore, Kwak et al. (2002) applied a multi-layer perceptron neural network to recognize the movement tracks of medaka (*Oryzias latipes*) in response to sub-lethal treatments of an insecticide.

3.2 Mathematical Framework

In this section, the mathematical background of the multi-layer perceptron neural networks is explained, including a review of background literature of successful implementations, an explanation of architectures, and a description of a method that was implemented to train the MLP.

A *neural network* is an information processing technique that is inspired by the way biological nervous systems, like the human brain, process information. It is a computer based machine, designed to model the way in which the brain performs a particular function of interest (Haykin 1999). It is an extraordinarily powerful mechanism that has found successful use in the diverse fields of mechanical engineering (Marwala and Hunt 1999; Vilakazi and Marwala 2007), civil engineering (Marwala 2000), aerospace engineering (Marwala 2001a; Marwala 2003), biomedical engineering (Marwala 2007), and finance (Patel and Marwala 2006). In this chapter, a multi-layer perceptron neural network is viewed as generalized regression model that can model both linear and non-linear data. The construction of a neural network involves four main steps (Marwala and Lagazio 2011; Msiza et al. 2007):

1. the processing units u_j, where each u_j has a certain activation level $a_j(t)$ at any point in time;
2. weighted inter-connections between a number of processing units. These inter-connections govern how the activation of one unit leads to the input for another unit;

3. an activation rule, which acts on the set of input signals at a processing unit to produce a new output signal; and
4. a learning rule that stipulates how to fine-tune the weights for a given input or output (Freeman and Skapura 1991).

Because they are capable of extracting meaning from complex data, neural networks are engaged to extract patterns and detect trends that are too complex to be identified by many other computer techniques (Hassoun 1995). A trained neural network can be viewed as an expert in the class of information that it has been given to analyze (Yoon and Peterson 1990). This expert can then be applied to offer predictions when presented with new circumstances. Because of their ability to adapt to non-linear data, neural networks have been applied to model a number of non-linear applications (Hassoun 1995; Leke et al. 2007).

The architecture of neural processing units and their inter-connections can have a significant influence on the processing capabilities of neural networks. Accordingly, there are many different connections that define how data flows between the input, hidden and output layers. The next section gives details on the architecture of the multi-layer perceptron neural networks employed in this chapter.

3.2.1 Multi-Layer Perceptrons (MLP) for Classification Problems

In the past 30 years the MLP neural networks have successfully been applied to both classification and regression problems. Ikuta et al. (2010) connected a chaos glial network to the MLP to solve a two-spiral problem and found that their method performed better than the conventional MLP. Zadeh et al. (2010) used the MLP to predict the daily flows from the Khosrow Shirin watershed, and observed that precipitation and discharge with a 1 day time lag best predicted daily flows. Narasinga-Rao et al. (2010) applied the multi-layer perceptron to predicting the quality of life in diabetes patients using age, gender, weight, and fasting plasma glucose as inputs.

Pasero et al. (2010) used the MLP for a time series analysis while Sug (2010) used a MLP for task classification. Zhang and Li (2009) used a hybrid of the Hidden Markov Model (HMM) and the MLP for speech recognition and demonstrated that the hybrid model performed better than the HMM.

He et al. (2009) used the MLP for short-term demand forecasting using graphics processing units. Bernardo-Torres and Gómez-Gil (2009) used the MLP to forecast seismograms while Sug (2009) applied the MLP to pilot sampling. Kushwaha and Shakya (2009) successfully applied the MLP for predicting the helical content of proteins while Karami et al. (2009) successfully applied it to decoding low-density parity-check codes.

Other applications for the MLP include the work by Achili et al. (2009) in robotics, Sancho-Gómez et al. (2009) for decision support systems with missing data, Krishna (2009) in an air data system, Hu and Weng (2009) in image processing, Duta and Duta (2009) in turbo-machinery optimization, Pontin et al. (2009) in predicting the occurrence of stinging jellyfish, Yazdanmehr et al. (2009) for the modeling of nanocrystals, Watts and Worner (2009) for predicting the distribution of fungal crop diseases as well as Yilmaz and Özer (2009) for pitch angle control in wind turbines above the rated wind speed.

In this book, neural networks are regarded as a broad structure for describing non-linear mappings between multi-dimensional spaces where the form of the mapping is overseen by a set of free parameters (Bishop 1995; Mohamed 2003; Mohamed et al. 2006) which have to be estimated from the data. There are two ways in which neural networks can be trained: supervised or unsupervised learning. We consider only supervised learning in this book. In supervised learning, the training data involves both the input to the neural networks and a corresponding target output. In this chapter, the input is a set of features that are deemed to influence the health status of the structure such as vibration data and the output is the identity of the faults.

3.2.2 Architecture

The multi-layer perceptron neural network architecture was selected for mapping the relationship between vibration data and the identity of faults. Each connection between inputs and neurons was weighted by adjustable weight parameters. In addition, each neuron had an adjustable bias weight parameter which is denoted by a connection from a constant input $x_0 = 1$ and $z_0 = 1$ for the hidden neurons and the output neuron, respectively. This group of two-layer multi-layer perceptron models can approximate any continuous function with arbitrary accuracy, as long as the number of hidden neurons is sufficiently large (Bishop 1995; Mohamed 2006).

The advantage of the multi-layer perceptron network is the interconnected cross-coupling that occurs between the input variables and the hidden nodes, with the hidden nodes and the output variables. If we assume that x is the input to the multi-layer perceptron and y is the output of the MLP, a mapping function between the input and the output may be written as follows (Bishop 1995):

$$y = f_{\text{output}}\left(\sum_{j=1}^{M} w_j f_{\text{hidden}}\left(\sum_{i=0}^{N} w_{ij} x_i\right) + w_0\right) \tag{3.1}$$

Here N is the number of inputs units, M is the number of hidden neurons, x_i is the ith input unit, w_{ij} is the weight parameter between input i and hidden neuron j and w_j is the weight parameter between hidden neuron j and the output neuron. The activation function $f_{\text{output}}(\cdot)$ is sigmoid and can be written as follows (Bishop 1995):

$$f_{output}(a) = \frac{1}{1 + e^{-a}} \tag{3.2}$$

In this chapter, we apply a neural network to classify the identity of fault. In modeling complex problems, care must be taken in the choice of the output activation function. For classification problems, as is the case in this chapter, the sigmoid function indicated in Eq. 3.2 is ideal (Bishop 1995). The activation function $f_{\text{hidden}}(\cdot)$ is a hyperbolic tangent which can be written as (Bishop 1995):

$$f_{hidden}(a) = \tanh(a) \tag{3.3}$$

3.2.3 Training of the Multi-Layer Perceptron

Once the activation function is defined and the sizes of the hidden nodes are selected, what remains is to approximate the network weights. The network weights are approximated from the observed data through a method called *training*. There are a number of critical matters that must be taken into account in training the networks. These include ensuring that, on identifying the network weights, the resulting network must not just memorize the data but learn from it. There are methods that have been presented to deal with this specific matter, and they comprise cross-validation, early-stopping, and regularization.

Two common methods can be applied to train a neural network. These methods are the *maximum likelihood technique* and the *Bayesian* method. The maximum likelihood technique estimates the network weights that maximize the capacity of a trained network to estimate the observed data, while the Bayesian method creates the probability distribution of the network model given the observed data. It should be noted here that the maximum likelihood and Bayesian methods are the same and the only difference is that the maximum likelihood technique identifies the network vector of weights that is most likely in the posterior probability distribution function. The first undertaking in identifying a vector of network weights that maximizes the predictive capacity of the neural network is to build a fitness function in the evolutionary programming perspective. The *fitness function* is a measure of the difference between the approximations of the model and the observed data.

In a two-class classification problem, the fitness function is the difference between the neural network's estimated output and the target output, t, given in Eq. 3.1 for all training patterns. E is the *cross-entropy error function* given by (Bishop 1995; Marwala 2009; Marwala and Lagazio 2011):

$$E = -\sum_{p=1}^{P} \{t_p \ln(y_p) + (1 - t_p)\ln(1 - y_p)\} \tag{3.4}$$

There are many advantages of the cross-entropy function. One of these includes the fact that it permits the output to be understood probabilistically without the necessity of invoking a Bayesian method. Neural networks are trained by iteratively adjusting the weight parameters, w, to minimize the fitness function given by Eq. 3.4. This is accomplished by randomly initializing the network weights and then adjusting the weight parameters using the scaled conjugate gradient technique (Møller 1993). The scaled conjugate gradient technique was selected over other optimization methods for the reason that it has efficient convergence properties.

Generalization is the capability of a trained neural network model to classify input patterns that were not observed during the training of the neural network. Essentially, on pursuing a network that generalizes, one identifies the balance between the capabilities of a network to remember the training data with the capability of the network to estimate data not seen. The generalization of performance is a true reflection of the capacity of a neural network to classify faults. This can simply be proven by separating the data into training and testing data sets.

Bishop (1995) demonstrated that a minimization of the cross-entropy fitness function in the neural network training with the activation function of a neural network results in the output of a neural network approximating the posterior probability of membership to a specific class, given the input $\{x\}$. In the present case of modeling the identity of faults, the output estimates the posterior probability of a specific identity of fault. If this class is represented by C_1 and the pattern class for not containing faults is represented by C_2, the relations for the posterior probability of class membership can be written as (Bishop 1995; Tettey and Marwala 2007):

$$P\left(C_1 \left|\{x\}\right.\right) = y \tag{3.5}$$

$$P\left(C_2 \left|\{x\}\right.\right) = 1 - y \tag{3.6}$$

Equations 3.5 and 3.6 offer a probabilistic interpretation to the neural network output. On the account of these relationships, it is obvious that the input vector has a high probability of being an element of class C_1 when y is close to 1 and C_2 when y is close to 0. If y is close to 0.5, then there is uncertainty in the class membership of the input vector. An elementary method to increase the efficacy of the classifier is to devise an upper and lower rejection threshold to the neural network output (Bishop 1995; Mohamed 2003). This classification decision rule can be expressed as follows (Mohamed 2003):

$$\begin{aligned} &\text{Choose } C_1 \text{ if } y > \gamma, \\ &\text{choose } C_2 \text{ if } y < (1-\gamma)\,, \\ &\text{otherwise do not classify } \{x\}. \end{aligned} \tag{3.7}$$

The parameter γ sets the level of the rejection threshold and permits the engineer to choose the level at which a decision can be made.

3.2.4 Back-Propagation Method

To identify the network weights given the training data, an optimization technique can be applied, within the context of the maximum-likelihood framework. In general, the weights can be identified using the following iterative technique (Werbos 1974):

$$\{w\}_{i+1} = \{w\}_i - \eta \frac{\partial E}{\partial \{w\}}(\{w\}_i) \tag{3.8}$$

In Eq. 3.8, the parameter η is the learning rate while $\{\}$ represents a vector. The minimization of the fitness function, *E*, is achieved by calculating the derivative of the errors, in Eq. 3.7, with respect to the network weight. The derivative of the error is calculated with respect to the weight which connects the hidden layer to the output layer, and can be written using the chain rule as follows (Bishop 1995; Marwala 2009):

$$\begin{aligned}\frac{\partial E}{\partial w_{kj}} &= \frac{\partial E}{\partial a_k}\frac{\partial a_k}{\partial w_{kj}}\\ &= \frac{\partial E}{\partial y_k}\frac{\partial y_k}{\partial a_k}\frac{\partial a_k}{\partial w_{kj}}\\ &= \sum_n f_{outer}'(a_k)\frac{\partial E}{\partial y_{nk}} z_j\end{aligned} \tag{3.9}$$

In Eq. 3.9, $z_j = f_{inner}(a_j)$ and $a_k = \sum_{j=0}^{M} w_{kj}^{(2)} y_j$. The derivative of the error with respect to weight which connects the hidden to the output layer may be written using the chain rule as (Bishop 1995):

$$\begin{aligned}\frac{\partial E}{\partial w_{kj}} &= \frac{\partial E}{\partial a_k}\frac{\partial a_k}{\partial w_{kj}}\\ &= \sum_n f_{inner}'(a_j)\sum_k w_{kj} f_{outer}'(a_k)\frac{\partial E}{\partial y_{nk}}\end{aligned} \tag{3.10}$$

In Eq. 3.10, $a_j = \sum_{i=1}^{d} w_{ji}^{(1)} x_i$. The derivative of the fitness function in Eq. 3.4 may thus be written as:

$$\frac{\partial E}{\partial y_{nk}} = \frac{t_{nk} - y_{nk}}{y_{nk}\left(y_{nk} - 1\right)} \tag{3.11}$$

while that of the hyperbolic tangent function is:

$$f'_{inner}(a_j) = \sec h^2(a_j). \tag{3.12}$$

The derivative of the logistic activation function is:

$$f'_{outer}(a_k) = f(a_k)(1 - f(a_k)) \tag{3.13}$$

Given that it has been determined how to compute the gradient of the error with respect to the network weights using back-propagation algorithms, Eq. 3.8 can be applied to update the network weights using an optimization process until some pre-defined stopping condition is achieved. If the learning rate in Eq. 3.8 is fixed then this is known as the *steepest descent optimization* method (Robbins and Monro 1951). Alternatively, since the steepest descent technique is not computationally efficient an improved technique needs to be developed, and in this chapter the scaled conjugate gradient method is implemented (Møller 1993), which is the concern of the next section.

3.2.5 Scaled Conjugate Gradient Method

The technique in which the network weights are inferred from the data is by applying some non-linear optimization technique (Mordecai 2003), and for this chapter is the scaled conjugate gradient technique. Before the scaled conjugate gradient technique is described, it is important to comprehend how it operates. As described before, the weight vector that gives the minimum error is attained by taking successive steps through the weight space as shown in Eq. 3.8 until some stopping criterion is achieved. Different algorithms select this learning rate differently. In this section, the gradient descent technique will be discussed, followed by how it can be extended to the *conjugate gradient technique* (Hestenes and Stiefel 1952). For the gradient descent technique, the step size is defined as $-\eta\partial E/\partial w$, where the parameter η is the learning rate and the gradient of the error is calculated using the back-propagation method described in the previous section.

If the learning rate is adequately small, the value of error decreases at each step until a minimum value is attained for the error between the model prediction and training target data. The disadvantage with this method is that it is computationally expensive when compared to other methods. For the conjugate gradient technique, the quadratic function of the error is minimized at each iteration over a progressively expanding linear vector space that includes the global minimum of the error (Luenberger 1984; Fletcher 1987; Bertsekas 1995).

For the conjugate gradient technique, the following steps are followed (Haykin 1999):

1. Choose the initial weight vector $\{w\}_0$.
2. Calculate the gradient vector $\frac{\partial E}{\partial\{w\}}(\{w\}_0)$.

3. At each step n use a line search to find the $\eta(n)$ that minimizes $E(\eta)$ which represents the cost function expressed in terms of η for fixed values of w and $-\frac{\partial E}{\partial\{w\}}(\{w_n\})$.
4. Check that the Euclidean norm of the vector $-\frac{\partial E}{\partial w}(\{w_n\})$ is sufficiently less than that of $-\frac{\partial E}{\partial w}(\{w_0\})$.
5. Update the weight vector using Eq. 3.8.
6. For w_{n+1} compute the updated gradient $\frac{\partial E}{\partial\{w\}}(\{w\}_{n+1})$.
7. Use the Polak-Ribiére method to calculate:

$$\beta(n+1) = \frac{\nabla E(\{w\}_{n+1})^T(\nabla E(\{w\}_{n+1}) - \nabla E(\{w\}_n)))}{\nabla E(\{w\}_n)^T \nabla E(\{w\}_n)}$$

8. Update the direction vector

$$\frac{\partial E}{\partial\{w\}}(\{w\}_{n+2}) = \frac{\partial E}{\partial\{w\}}(\{w\}_{n+1}) - \beta(n+1)\frac{\partial E}{\partial\{w\}}(\{w\}_n).$$

9. Set $n = n + 1$ and go back to step 3.
10. Stop when the following condition is satisfied: $\frac{\partial E}{\partial\{w\}}(\{w\}_{n+2}) = \varepsilon\frac{\partial E}{\partial\{w\}}(\{w\}_{n+1})$ where ε is a small number.

The scaled conjugate gradient method differs from the conjugate gradient method in that it does not involve the line search explained in step 3. The step-size (see step 3) can be calculated directly by using the following formula (Møller 1993):

$$\eta(n) = 2\left(\eta(n) - \frac{\left(\frac{\partial E(n)}{\partial\{w\}}(n)\right)^T H(n)\left(\frac{\partial E(n)}{\partial\{w\}}(n)\right)\ldots}{+\,\eta(n)\left\|\left(\frac{\partial E(n)}{\partial\{w\}}(n)\right)\right\|^2} \Big/ \left\|\left(\frac{\partial E(n)}{\partial\{w\}}(n)\right)\right\|\right)^2 \tag{3.14}$$

Here H is the Hessian of the gradient. The scaled conjugate gradient method was used because it has been found to solve the optimization problems encountered when training an MLP network to be more computationally efficient than the gradient descent and conjugate gradient methods (Bishop 1995).

3.3 The Multifold Cross-Validation Method

In the example of cylindrical shells considered in this chapter and described in Chap. 2 (Marwala 2001b), because there is a limited amount of data available, the training data set was also applied as a validation data set by using a multifold cross-validation

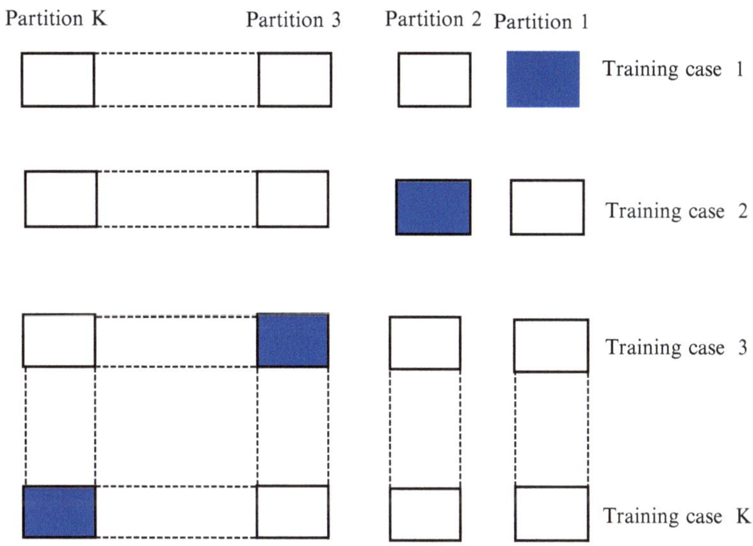

Fig. 3.1 The multifold cross-validation technique applied where the network was trained K times, each time leaving out the data indicated by the shaded area and using the omitted data for validation purposes. The validation error was acquired by averaging the squared error under validation over all the trials of the experiment

technique (Stone 1974; Kohavi 1995). The multifold method implemented in the present study is illustrated in Fig. 3.1 (Marwala 2001b). Each column in Fig. 3.1 demonstrates a partition of the training data set and each row denotes a training case. The shaded box for a given training case is the partition that is applied for validation purposes whereas the rest of the boxes in one row are applied to train the network.

When the multifold cross-validation method is applied, the training data set with N examples is segmented into K partitions. Here it is assumed that N is divisible by K and that $K > 1$. For each training process, the network is trained with the data from all partitions excluding one and the validation set is the subset that is left out. The partition that is left out for each training case is a shaded one. For example, in for Training case 1 the network is trained using Partitions 2 to K and Partition 1 is used as a validation set. The procedure is repeated for K training cases, by leaving the shaded partition for validation and using the remaining partitions for training. It should be noted that the type of the multifold cross-validation technique applied in this chapter resets the network once in Training case 1. The network-weights attained after Training case 1 turn into initial network weights for Training case 2 and so on. The performance of the resulting network is assessed by averaging the mean squared errors or classification error over all the training cases.

If the amount of data is inadequate or stark, then a technique called the *leave-one-out process,* was applied, which is a distinct case of the multifold cross-validation technique, where all examples but one are used to train the network and the model is validated on the remaining one. The research directed by Shao and Tu (1995) advises that the multifold cross-validation technique performs better than the leave-one-out scheme for approximating generalization errors. This is because the leave-one-out technique over-fits the data. For each training session there must be a stopping criterion. For this book, training was stopped after 50 scaled conjugate gradient iterations had lapsed.

3.4 Application to Cylindrical Shells

For this chapter the multifold cross-validation technique was used to train and validate the pseudo-modal-energy-network and modal-property-network. A pseudomodal-energy-network is a Multi-Layer Perceptron neural network that was trained using pseudo-modal energies while a modal-property-network uses modal properties which were described in Chap. 2. The fault cases that were used to train and test the networks are shown in Table 3.1 (Marwala 2001b).

In Table 3.1 the training data set, with 168 fault cases, has an equal number of fault cases showing that the probabilities of incidence for the eight fault cases are equal. The remaining 96 fault cases were used to test the networks. The training data set with 168 fault cases was subdivided into 21 subsets. Each partition had eight different fault cases. This established that the training set was balanced in terms of the proportion of fault cases present. The first sets of networks, *i.e.,* the pseudo-modal-energy-network and the modal-property-network (20 for each method), were trained with 160 fault cases (from Partitions 2 to 21) and the networks were validated on the remaining eight fault cases (from Partition 1). The network weights identified were used as initial weights for Training case 2. The training for this case was conducted using all partitions apart from Partition 2, which was used to validate the trained networks. The complete training and validation of the networks were conducted 21 times until all the validation partitions had been used. As already revealed, 20 pseudo-modal-energies with the number of hidden units randomly chosen to fall between 7 and 11 were trained and validated using the multifold cross-validation process. The same process was used to train 20 modal-property-networks. From these two sets of 20 trained networks, the pseudo-modal-energy-network and modal-property-network that gave the least mean squared errors over the validation partitions were selected. Each validation partition gave a mean squared error. The average of the mean squared errors of all the partitions was the validation error was used to select the networks. The pseudo-modal-energy-network and modal-property-network that had the least mean squared errors had respectively 10 inputs

Table 3.1 Fault cases used to train, cross-validate and test the networks

Fault	[000]	[100]	[010]	[001]	[110]	[101]	[011]	[111]
Training set	21	21	21	21	21	21	21	21
Test set	39	3	3	3	3	3	3	39

Table 3.2 Confusion matrix from the classification of fault cases in the test data using the pseudo-modal-energy-network

		Predicted							
		[000]	[100]	[010]	[001]	[110]	[101]	[011]	[111]
Actual	[000]	36	0	2	1	0	0	0	0
	[100]	0	3	0	0	0	0	0	0
	[010]	0	0	3	0	0	0	0	0
	[001]	0	0	0	3	0	0	0	0
	[110]	0	0	0	0	3	0	0	0
	[101]	0	0	0	0	0	3	0	0
	[011]	0	0	0	0	0	0	3	0
	[111]	0	0	1	0	4	3	6	25

Table 3.3 Confusion matrix from the classification of fault cases in the test data using the modal-property-network

		Predicted							
		[000]	[100]	[010]	[001]	[110]	[101]	[011]	[111]
Actual	[000]	35	0	1	3	0	0	0	0
	[100]	0	3	0	0	0	0	0	0
	[010]	0	0	3	0	0	0	0	0
	[001]	0	0	0	3	0	0	0	0
	[110]	0	0	0	0	3	0	0	0
	[101]	0	0	0	3	0	0	0	0
	[011]	0	0	0	0	0	0	3	0
	[111]	0	0	0	0	8	1	2	28

and 8 hidden nodes as well as 10 inputs and 9 hidden nodes. The number of input units was chosen. Here 10 input units were selected using the principal component analysis as explained by Marwala (2001b).

Fault cases given by a network were rounded off to the nearest whole number, i.e., 0 and 1. To assess the predictive capabilities of the trained set of networks, a confusion matrix was applied as shown in Table 3.2. In this table the predicted fault cases are displayed vertically and the actual fault cases are displayed horizontally. A row of this matrix indicates all fault cases present in the test data for that particular fault case. As an example, a row with a fault case [000] in Table 3.2 signifies the number of [000] fault cases used in the test data set. From the confusion matrix certain information may be derived. The diagonal components of this matrix demonstrate fault cases classified correctly, while the off-diagonal components of this matrix represent fault cases classified wrongly. A perfect fault identification procedure indicates a diagonal matrix with all off-diagonal components equal to zero. A completely imperfect confusion matrix indicates zero diagonal components and non-zero off-diagonal components. The results showing the confusion matrices when the pseudo-modal-energy-network and modal-property-network were used, are given in Tables 3.2 and 3.3 respectively (Marwala 2001b).

In Table 3.2, 92.3% of [000] cases; all the one- and two-fault cases; and 64.1% of [111] cases were correctly classified. Of the three [000] fault cases that were classified wrongly using the pseudo-modal-energy-network, two were classified as [010] cases and one as a [001] case. Of the fourteen [111] cases that were classified wrongly by the pseudo-modal-energy-network, four were classified as [110] cases, three as [101] cases, six as [011] cases, and one as a [010] case.

The confusion matrix obtained when the modal-property-network was used is shown in Table 3.3 (Marwala 2001b). This table demonstrates that this network classifies 89.7% of [000] fault cases correctly; all one and two-fault cases with, the exception of three [101] cases correctly; and 71.8% of [111] fault cases correctly. Of the four [000] cases that were classified wrongly by the modal-property-network, one is classified as a [010] case and three as [001] cases. Of the eleven [111] cases that were classified wrongly by the modal-property-network, eight were classified as [110] cases, one as a [101] case and two as [011] cases. The three [101] cases that were misclassified by the modal-property-network were all classified wrongly as [001] cases.

The pseudo-modal-energy-network misclassified three cases and the modal-property-network misclassified four [000] cases. The pseudo-modal-energy-network classified all the one- and two- fault cases correctly, while the modal-property-network misclassified all [101] cases. The modal-property-network misclassified eleven [111] cases and the pseudo-modal-energy-network misclassified fourteen [111] cases.

The results indicate that in classifying all fault cases, the pseudo-modal-energy-network was only marginally better than the modal-property-network. Nevertheless, if account is taken of the fact that the modal-property-network could not correctly classify an entire fault case, *i.e.,* [101], where this was never the case for the pseudo-modal-energy-network, then it can be concluded that the pseudo-modal-energy-network was better than the modal-property-network.

3.5 Conclusion

In this chapter, modal properties and pseudo-modal energies data as well as the multi-layer perceptron network were applied to classify faults in a population of cylindrical shells and were experimentally validated. A principal component analysis was applied to reduce the dimensions of the input data. The multifold cross validation technique was applied to choose the optimal number of hidden units amongst the 20 trained pseudo-modal-energy-networks and the 20 trained modal-property-networks. The pseudo-modal-energy-network and the modal-property-network were found to offer similar levels of accuracy on classifying faults.

References

Abu-Mahfouz I (2003) Drilling wear detection and classification using vibration signals and artificial neural network. Int J Mach Tools Manufacture 43(7):707–720

Achili B, Daachi B, Ali-Cherif A, Amirat Y (2009) Combined multi-layer perceptron neural network and sliding mode technique for parallel robots control: an adaptive approach. In: Proceedings of the international joint conference on neural network, Atlanta, Georgia, pp 28–35

Bernardo-Torres A, Gómez-Gil P (2009) One-step forecasting of seismograms using multi-layer perceptrons. In: Proceedings of the 6th international conference on electrical engineering, computer science and automatic control, Athens, Greece, pp 1–4

Bertsekas DP (1995) Non-linear programming. Athenas Scientific, Belmont

Bishop CM (1995) Neural networks for pattern recognition. Oxford University Press, Oxford

Bucolo M, Fortuna L, Nelke M, Rizzo A, Sciacca T (2002) Prediction models for the corrosion phenomena in pulp & paper plant. Control Eng Pract 10(2):227–237

Caputo AC, Pelagagge PM (2002) An inverse approach for piping networks monitoring. J Loss Prev Process Ind 15(6):497–505

Dimla DE, Lister PM (2000) On-line metal cutting tool condition monitoring: II: tool-state classification using multi-layer perceptron neural networks. Int J Mach Tools Manufacture 40(5):769–781

Duta MC, Duta MD (2009) Multi-objective turbomachinery optimization using a gradient-enhanced multi-layer perceptron. Int J Numer Methods Fluids 61:591–605

Fletcher R (1987) Practical methods of optimization, 2nd edn. Wiley, New York

Freeman J, Skapura D (1991) Neural networks: algorithms, applications and programming techniques. Addison-Wesley, Reading

Hassoun MH (1995) Fundamentals of artificial neural networks. MIT Press, Cambridge

Haykin S (1999) Neural networks. Prentice-Hall, Upper Saddle River

He T, Dong ZY, Meng K, Wang H, Oh YT (2009) Accelerating multi-layer perceptron based short-term demand forecasting using graphics processing units. In: Transmision & distribution conference & exposition: Asia and Pacific, Seoul, Korea, pp 1–4

Herzog MA, Marwala T, Heyns PS (2009) Machine and component residual life estimation through the application of neural networks. Reliability Eng Syst Saf 94(2):479–489

Hestenes MR, Stiefel E (1952) Methods of conjugate gradients for solving linear systems. J Res Natl Bur Stand 6:409–436

Hu X, Weng Q (2009) Estimating impervious surfaces from medium spatial resolution imagery using the self-organizing map and multi-layer perceptron neural networks. Remote Sens Environ 113:2089–2102

Ikuta C, Uwate Y, Nishio Y (2010) Chaos glial network connected to multi-layer perceptron for solving two-spiral problem. In: Proceedings of IEEE international symposium on circuits and systems: nano-bio circuit fabrics and systems, Paris, France, pp 1360–1363

Karami AR, Ahmadian-Attari M, Tavakoli H (2009) Multi-layer perceptron neural networks decoder for LDPC codes. In: Proceedings of the 5th international conference on wireless communications, networking and mobile computing, pp 1–4

Kim C-K, Kwak I-S, Cha E-Y, Chon T-S (2006) Implementation of wavelets and artificial neural networks to detection of toxic response behavior of chironomids (Chironomidae: Diptera) for water quality monitoring. Ecol Model 195(1–2): 61–71, Selected Papers from the third conference of the International Society for Ecological Informatics (ISEI), 26–30 August, 2002, Grottaferrata, Rome, Italy

Kohavi R (1995) A study of cross-validation and bootstrap for accuracy estimation and model selection. In: Proceedings of the 4th international joint conference on artificial intelligence, pp 1137–1143

Krishna HS (2009) Highly accurate multi-layer perceptron neural network for air data system. Def Sci J 59:670–674

Kushwaha SK, Shakya M (2009) Multi-layer perceptron architecture for tertiary structure prediction of helical content of proteins from peptide sequences. In: Proceedings of the international conference on advances in recent technologies in communication and computing, pp 465–467

Kwak I-S, Chon T-S, Kang H-M, Chung N-I, Kim J-S, Koh SC, Lee S-K, Kim Y-S (2002) Pattern recognition of the movement tracks of medaka (*Oryzias latipes*) in response to sub-lethal treatments of an insecticide by using artificial neural networks. Environ Pollut 120(3):671–681

Leke B, Marwala T, Tettey T (2007) Using inverse neural network for HIV adaptive control. Int J Comput Intell Res 3:11–15

Luenberger DG (1984) Linear and non-linear programming, 2nd edn. Addison-Wesley, Reading

Marwala T (2000) On damage identification using a committee of neural networks. J Eng Mech 126:43–50

Marwala T (2001a) Probabilistic fault identification using a committee of neural networks and vibration data. J Aircraft 38:138–146

Marwala T (2001b) Fault identification using neural networks and vibration data. Doctor of Philosophy Topic, University of Cambridge, Cambridge, UK

Marwala T (2003) Fault classification using pseudo modal energies and neural networks. Am Inst Aeronaut Astronaut J 41:82–89

Marwala T (2007) Bayesian training of neural network using genetic programming. Pattern Recognit Lett 28:1452–1458

Marwala T (2009) Computational intelligence for missing data imputation, estimation and management: knowledge optimization techniques. IGI Global Publications, New York

Marwala T, Hunt HEM (1999) Fault identification using finite element models and neural networks. Mech Syst Signal Process 13:475–490

Marwala T, Lagazio M (2011) Militarized conflict modeling using computational intelligence techniques. Springer, London

Mohamed N (2003) Detection of epileptic activity in the EEG using artificial neural networks. M.Sc. (Electrical Engineering) Thesis, University of the Witwatersrand

Mohamed N, Rubin D, Marwala T (2006) Detection of epileptiform activity in human EEG signals using Bayesian neural networks. Neural Inf Process Lett Rev 10:1–10

Mohamed S (2006) Dynamic protein classification: adaptive models based on incremental learning strategies. Unpublished Master's thesis, University of the Witwatersrand, Johannesburg

Møller MF (1993) A scaled conjugate gradient algorithm for fast supervised learning. Neural Netw 6:525–533

Mordecai A (2003) Non-linear programming: analysis and methods. Dover Publishing, New York, USA

Msiza IS, Nelwamondo FV, Marwala T (2007) Water demand forecasting using multi-layer perceptron and radial basis functions. In: Proceedings of the IEEE international conference on neural networks, Orlando, Florida, pp 13–18

Mustapha F, Manson G, Worden K, Pierce SG (2007) Damage location in an isotropic plate using a vector of novelty indices. Mech Syst Signal Process 21(4):1885–1906

Narasinga-Rao MR, Sridhar GR, Madhu K, Rao AA (2010) A clinical decision support system using multi-layer perceptron neural network to predict quality of life in diabetes. Diabetes Metab Syndr: Clin Res Rev 4:57–59

Pasero E, Raimondo G, Ruffa S (2010) MULP: a multi-layer perceptron application to long-term, out-of-sample time series prediction. Lect Notes Comput Sci LNCS6064:566–575

Patel P, Marwala T (2006) Neural networks, fuzzy inference systems and adaptive-neuro fuzzy inference systems for financial decision making. Lect Notes Comput Sci LNCS4234:430–439

Pontin DR, Worner SP, Watts MJ (2009) Using time lagged input data to improve prediction of stinging jellyfish occurrence at New Zealand beaches by multi-layer perceptrons. Lect Notes Comput Sci LNCS5506:909–916

Rafiee J, Arvani F, Harifi A, Sadeghi MH (2007) Intelligent condition monitoring of a gearbox using artificial neural network. Mech Syst Signal Process 21(4):1746–1754

Robbins H, Monro S (1951) A stochastic approximation method. Ann Math Stat 22:400–407

Sancho-Gómez JL, García-Laencina PJ, Figueiras-Vidal AR (2009) Combining missing data imputation and pattern classification in a multi-layer perceptron. Intell Autom Soft Comput 15:539–553

Shao J, Tu D (1995) The jackknife and bootstrap. Springer, New York

Stone M (1974) Cross-validatory choice and assessment of statistical predictions. J Stat Soc B36:111–113

Sug H (2009) A pilot sampling method for multi-layer perceptrons. In: Proceedings of the 13th WSEAS international conference on computers, Athens, Greece, pp 629–633

Sug H (2010) Investigating better multi-layer perceptrons for the task of classification. WSEAS Trans Comput 9:475–485

Tettey T, Marwala T (2007) Conflict modelling and knowledge extraction using computational intelligence methods. In: Proceedings of thc 11th international conference on intelligent engineering systems, pp 161–166

Vilakazi BC, Marwala T (2007) Condition monitoring using computational intelligence. In: Laha D, Mandal P (eds) Handbook on computational intelligence in manufacturing and production management, illustrated edn. IGI Publishers, New York

Watts MJ, Worner SP (2009) Predicting the distribution of fungal crop diseases from abiotic and biotic factors using multi-layer perceptrons. Lect Notes Comput Sci LNCS5506:901–908

Werbos PJ (1974) Beyond regression: new tool for prediction and analysis in the behavioral sciences. Doctoral dissertation, Harvard University

Yazdanmehr M, Anijdan SHM, Samadi A, Bahrami A (2009) Mechanical behavior modeling of nanocrystalline NiAl compound by a feed-forward back-propagation multi-layer perceptron ANN. Comput Mater Sci 44:1231–1235

Yella S, Dougherty M, Gupta NK (2009) Condition monitoring of wooden railway sleepers. Transportation Res C: Emerg Technol 17(1):38–55

Yilmaz AS, Özer Z (2009) Pitch angle control in wind turbines above the rated wind speed by multi-layer perceptron and radial basis function neural networks. Expert Syst Appl 36:9767–9775

Yoon Y, Peterson LL (1990) Artificial neural networks: an emerging new technique. In: Proceedings of the ACM SIGBDP conference on trends and direction in expert systems, Orlando, Florida, pp 7–422

Zadeh MR, Amin S, Khalili D, Singh VP (2010) Daily outflow prediction by multi layer perceptron with logistic sigmoid and tangent sigmoid activation functions. Water Res Manag 24:2673–2688

Zhang P, Li H (2009) Hybrid model of continuous hidden Markov model and multi-layer perceptron in speech recognition. In: Proceedings of the 2nd international conference on intelligent computation technology and automation, Guangdong, China, pp 62–65

Chapter 4
Bayesian Approaches to Condition Monitoring

4.1 Introduction

For this chapter, Bayesian networks were trained using the hybrid Monte Carlo method with vibration data used for the application of monitoring the condition of cylindrical shells. Weidl et al. (2005) applied generic object-oriented Bayesian networks for condition monitoring, root-cause analysis and decision support to the operation of complex continuous processes. Their technique combined a decision-theoretic diagnostic with risk assessment of an industrial process control and was implemented for a pulp digesting and screening process. Their results showed that the system did perform reasoning under uncertainty and offered remedial actions to the operators with reasons for the root causes and detailed operators' activities for related examples. The Bayesian network models were arranged to execute sequential learning to increase its diagnostic performance.

Kohda and Cui (2005) applied Bayesian network for the risk-based reconfiguration of a safety monitoring system to avoid an atypical episode from progressing to an accident. Their safety monitoring system detected indicators of unusual incidents and mitigated its influence. A case study of a three-sensor system demonstrated the advantages of their technique.

Marwala (2007) applied Bayesian techniques for training of neural networks using genetic programming. Bayesian neural network were trained using Markov Chain Monte Carlo (MCMC) and genetic programming in binary space within a Metropolis framework. The procedure proposed had the ability to learn using samples obtained from previous steps merged using concepts of natural evolution which included mutation, crossover and reproduction. The reproduction function was the Metropolis framework; binary mutation and simple crossover were also used. Their algorithm was tested for condition monitoring of structures and the results were compared to those of a classical MCMC method. The results confirmed that Bayesian neural networks trained using genetic programming offered a better performance and efficiency than the classical approach.

T. Marwala, *Condition Monitoring Using Computational Intelligence Methods: Applications in Mechanical and Electrical Systems*, DOI 10.1007/978-1-4471-2380-4_4,

Qi and Huang (2011) successfully applied Bayesian techniques for control loop diagnosis in the presence of temporal dependent evidences. Traditional Bayesian approaches usually assume that evidences are temporally independent but this condition does not apply in many engineering cases. By assuming that the evidence transition information needs to be considered, the temporal information can be derived within the Bayesian framework to advance diagnosis performance. Qi and Huang (2011) solved the evidence dependency case by applying a data-driven Bayesian method with attention to the evidence transition probability.

Katsis et al. (2011) applied a wearable system for the affective monitoring of car racing drivers during simulated conditions. The wearable device was intended to gather chosen biological signals, pre-process them and wirelessly transmit them from the site of the subject to the centralized system. The centralized system was intended to conduct an assessment of the subject's emotional state and project a generic 3D face model where the facial expression of the subject could be observed. A two stage classification system was used. The system entailed a decision tree to classify the subject's emotional state as high stress, low stress and valence as well as a Tree Augmented Naive Bayesian to classify into two classes: euphoria and dysphoria. The system was validated using a dataset obtained from ten subjects in simulated racing conditions and the overall classification rate achieved using tenfold cross-validation was very good.

Droguett et al. (2008) applied a semi-Markov model with a Bayesian belief network based human error probability for assessing the availability of down-hole optical monitoring of oil fields in Brazil. They developed a pressure-temperature optical monitoring systems by using an availability assessment model where system dynamics were described using a continuous-time semi-Markovian process quantified using probabilities, which was combined with a Bayesian belief network describing the cause-effect relationships among features influencing the repairman's error probability during maintenance.

Kim et al. (2011) successfully applied a Bayesian framework to identify faults in a partially observable system subject to random failures. The deterioration of a system was modeled with a hidden 3-state continuous time homogeneous Markov process with States 0 and 1 not being observable, signifying good and warning situations respectively, and only State 2 being observable. The model's parameters were identified using an Expectation Maximization procedure and a cost-optimal Bayesian fault prediction scheme was presented. The technique was validated using real data from a spectrometric analysis of oil samples from the transmission units of heavy hauler trucks.

Yuen and Kuok (2010) studied the long-term monitoring of a 22-storey reinforced concrete building. It is necessary to distinguish the inevitable variations in ambient conditions resulting from the abnormal changes due to structural damage and deterioration. A Bayesian framework was implemented to quantify the uncertain parameters in the modal frequency-ambient condition model. The results showed that direct attention to the ambient temperature and relative humidity was vital for long-term structural health monitoring.

Pollino et al. (2007) investigated conflicts and improved strategies for the management of an endangered Eucalypt species using Bayesian networks. Their framework gave a procedure to guide future integrative and iterative monitoring and research.

Subrahmanya et al. (2010) applied a Bayesian machine learning technique for sensor choice and fusion in an on-board fault diagnostics process. They presented a procedure for choosing groups of features during regression and classification. A hierarchical Bayesian framework introduced grouping for the parameters of a generalized linear model and the model's hyper-parameters were approximated using an empirical Bayes procedure. The performance of the procedure was first tested on a synthetic regression example and applied to fault detection in diesel engines.

Willis (2010) applied a logistic Partial Least Squares method for the condition monitoring of vibrations in the centrifuge of a product treatment plant. A logistic Partial Least Squares model was obtained using wavelet coefficients to de-correlate the time series data. The model offered a reference line to assess any development through kernel techniques. The kernel hypothesis was presented from a Bayesian viewpoint and applied to create a detector with considerably fewer false positives and missed detections.

Nebot et al. (2007) applied Bayesian networks for the diagnosis of surface roughness and cutting tool-wear. They presented a multi-sensor system for indirect monitoring. Their rationale for using a Bayesian network was its ability to handle the stochastic characteristics of the machining process. Their results were that models with a high discretization showed a reliability of 89.5% for the surface roughness prediction and 97.3% for the cutting tool wear diagnosis, whereas lower discretizations gave better reliability but worse diagnosis.

Feng and Schlindwein (2009) applied normalized wavelet packets quantifiers for tool condition monitoring. They used Acoustic Emission signals from faulty bearings of rotating machines and demonstrated that localized defects and advanced contamination faults can be successfully identified if a suitable quantifier was selected. When the Bayesian classifier was applied to quantitatively analyze and evaluate the performance of their quantifiers it was shown that decreasing the Daubechies wavelet order or the length of the segment deteriorated the performance of the quantifiers.

Hu et al. (2010) applied an integrated technique for safety pre-warning in complex large-scale and industrial systems. They implemented an integrated technique that combined degradation process modeling, a dynamic Bayesian network, condition monitoring, safety assessment and prognosis steps, taking into account the *a priori* knowledge of the interactions and dependencies among components and the environment, the relationships between hazard causes and effects, and historical failure data and online real-time data from condition monitoring. Their application of the integrated safety pre-warning method to the gas turbine compressor system revealed how each phase of the proposed technique contributed to the accomplishment of the development of a safety pre-warning system in a systematic manner.

Other successful applications of a Bayesian framework include studies in clinical trials by Daimon (2008), the dynamics of attentional control under conflict (Yu et al. 2009), and in the analysis of plant colonization on an Arctic moraine since the end of the Little Ice Age (Moreau et al. 2005).

The next section describes the neural network which was used for the fault identification of this chapter.

4.2 Neural Networks

For this chapter, as for the previous chapter, multi-layer perceptron neural network models were expressed in the Bayesian context and trained through applying Monte Carlo methods (Marwala 2001; Marwala and Lagazio 2011). These techniques were implemented for the classification of faults in a population of cylindrical shells. Hence this section gives a summary of neural networks within the context of fault classification problems. For this chapter, a multi-layer perceptron was applied to map the modal properties and pseudo-modal energies (x) and the fault classification in a population of cylindrical shells (y). The relationship between the kth identity of fault, y_k, and the pseudo-modal energies or modal properties, x, may be written as follows (Bishop 1995; Marwala 2009; Marwala and Lagazio 2011):

$$y_k = f_{outer}\left(\sum_{j=1}^{M} w_{kj}^{(2)} f_{inner}\left(\sum_{i=1}^{d} w_{ji}^{(1)} x_i + w_{j0}^{(1)}\right) + w_{k0}^{(2)}\right) \tag{4.1}$$

Here $w_{ji}^{(1)}$ and $w_{kj}^{(2)}$ indicate the weights in the first and second layers, respectively, going from input i to hidden unit j, M is the number of hidden units, d is the number of output units, while $w_{j0}^{(1)}$ indicates the bias for the hidden unit j and $w_{k0}^{(2)}$ indicates the bias for the output unit k.

Choosing an appropriate network architecture is an important prerequisite for model construction. For this chapter, the architecture chosen was the Multi-Layered Perceptron (MLP), and in Chap. 3 was trained by applying the scaled conjugate gradient method (Moller 1993). When choosing an appropriate MLP model, another important decision lies in the choice of the correct number of hidden units (M), and the class of functional transformations that they accomplish. This is because a large value of M will produce very flexible networks, which may learn not only the data configuration but also the noise in the data. Conversely, a small value of M will produce networks that are unable to model complex relationships. To identify the optimal MLP structure, the network was trained various times through applying the scaled conjugate gradient technique. The problem of identifying the weights and biases in neural networks may be posed in the Bayesian framework as (MacKay 1991; Bishop 1995; Lagazio and Marwala 2005; Marwala 2009; Marwala and Lagazio 2011):

$$P(w|[D]) = \frac{P([D]|w)P(w)}{P([D])} \tag{4.2}$$

Here $P(w)$ is the probability distribution function of the weight-space in the absence of any data, also called the *prior distribution function* and $[D]\equiv(y_1, \ldots, y_N)$ is a matrix containing the identity of fault data. The expression $P(w|[D])$ is the posterior probability distribution function after the data have been observed, $P([D|w])$ is the likelihood function and $P([D])$ is the normalization function, also known as the "evidence". For the MLP, Eq. 4.2 may be expanded by applying the cross-entropy error function to give (MacKay 1992; Bishop 1995; Marwala 2009; Marwala and Lagazio 2011):

$$P(w|[D]) = \frac{1}{Z_s} \exp\left(\beta \sum_n^N \sum_k^K \{t_{nk} \ln(y_{nk}) + (1 - t_{nk}) \ln(1 - y_{nk})\} - \sum_j^W \frac{\alpha_j}{2} w_j^2\right) \tag{4.3}$$

where

$$Z_S(\alpha, \beta) = \left(\frac{2\pi}{\beta}\right)^{N/2} + \left(\frac{2\pi}{\alpha}\right)^{W/2} \tag{4.4}$$

The cost-entropy error function was applied because of its classification advantages. Also a weight-decay was assumed for the prior distribution as it penalizes the weights with large magnitudes. In Eq. 4.3, n is the index for the training pattern, hyper-parameter β is the data contribution to the error, k is the index for the output units, t_{nk} is the target output corresponding to the nth training pattern and k^{th} output unit and y_{nk} is the corresponding predicted output. The parameter α_j is another hyper-parameter, which determines the relative contribution of the regularization term on the training error. In Eq. 4.3, the hyper-parameters may be set for groups of weights. Equation 4.3 can be solved in two ways: by using the Taylor expansion and through approximating it as a Gaussian distribution and applying the evidence framework; or by numerically sampling the posterior probability by applying approaches such as the Monte Carlo technique, simulated annealing, the genetic Monte Carlo method, or the hybrid Monte Carlo (Marwala 2010). The following section describes some of these sampling approaches.

4.3 Sampling Methods

In turn, this section describes the following sampling methods: the Monte Carlo Method, the Markov Chain Monte Carlo Method, the hybrid Monte Carlo method. These methods were also described in detail by Marwala and Lagazio (2011).

4.3.1 Monte Carlo Method

Monte Carlo approaches are regularly used to simulate complex systems. Monte Carlo approaches are a type of numerical method that depends on repetitive random sampling to approximate the results. Due to their dependence on recurrent computation of random or simulated random numbers, these methods are well suited for approximate results using computers and are used when it is unrealistic to approximate a solution using a deterministic method (Marwala and Lagazio 2011).

The Monte Carlo technique is a computational procedure that applies recurrent random sampling to compute a result (Mathe and Novak 2007; Akhmatskaya et al. 2009; Ratick and Schwarz 2009: Marwala 2009, 2010; Marwala and Lagazio 2011). Monte Carlo techniques have been applied for simulating physical and mathematical systems. For instance, Lai (2009) applied the Monte Carlo technique to solving matrix and integral problems while McClarren and Urbatsch (2009) applied an adapted Monte Carlo technique for modeling time-dependent radiative transfer with adaptive material coupling.

Other recent applications of the Monte Carlo technique include its use in particle coagulation (Zhao and Zheng 2009), in diffusion problems (Liu et al. 2009), for the design of radiation detectors (Dunn and Shultis 2009), for modeling bacterial activities (Oliveira et al. 2009), for vehicle detection (Jia and Zhang 2009), for modeling the bystander effect (Xia et al. 2009), and for modeling nitrogen absorption (Rahmati and Modarress 2009).

Kandela et al. (2010) applied the Monte Carlo technique to study the movement of tablets in a pan coater by using video imaging. They applied the technique to track the motion of tablets and used coating variables of circulation time, surface time, projected surface area and surface velocity of the tablet. These parameters were derived from video imaging experiments. Other applications of the Monte Carlo technique include Padilla Cabal et al. (2010) who applied the technique to approximate the efficiency of an n-type HPGe detector as well as Fefelov et al. (2009) who applied the Monte Carlo technique to study a self-assembled monolayer with a number of different orientations for the organic molecules.

Martin and Ayesa (2010) applied the Monte Carlo technique to calibrate water quality models while Roskilly et al. (2010) applied it to examine the effect of shape on particle separation. Do et al. (2010) applied Monte Carlo techniques to simulate the vapor–liquid equilibrium properties of R134a and its liquid microscopic structure and observed that the simulations agreed with experimental data. Ozaki et al. (2010) applied the Monte Carlo technique to develop a framework for data analysis, including a process to link and control data processing modules.

Monte Carlo simulation approaches are advantageous in analyzing systems with a large number of degrees of freedom and uncertain inputs in varied fields such as fluid dynamics, materials science, and solid mechanisms (Robert and Casella

2004). The Monte Carlo technique normally follows the following practice (Robert and Casella 2004; Marwala and Lagazio 2011):

- Express the input space.
- Randomly create inputs from the input space by applying a selected probability distribution.
- Apply the produced input for the deterministic calculation.
- Integrate the results of the individual calculations to approximate the final result.

A simple example that has been used many times to explain the Monte Carlo technique is the approximation of π by drawing a square and putting a circle inside it. The area of the square is $4r^2$ while the area of the circle inside it is πr^2. The ratio of the area of the circle to the area of the square is $\pi/4$. By applying the Monte Carlo technique, the input space is any point inside the square. If data points are randomly produced to be located inside the square, the ratio of the number of points that are located inside the circle to the ratio of the points that are located inside the square is equal to $\pi/4$. This way, the value of π can be approximated experimentally.

4.3.2 Markov Chain Monte Carlo Method

An additional method of sampling the posterior probability is to use the Markov Chain Monte Carlo (MCMC) technique, which is a random walk Monte Carlo routine which is performed through generating a Markov chain to identify an equilibrium distribution. The MCMC comprises of a Markov process and a Monte Carlo simulation (Liesenfeld and Richard 2008). After many random walk steps, the retained states will converge to a desired posterior distribution. Technically, as the number of steps approach infinity, the accuracy of the estimated probability distribution becomes ideal. Rodina et al. (2010) applied the MCMC to predict renal disease, while Drugan and Thierens (2010) presented an evolutionary MCMC where evolutionary methods were applied to exchange information between states. Wang et al. (2010) applied the MCMC for spectrum sensing in cognitive radio while Wang and Harrison (2010) applied the MCMC to describe a water distribution system.

Wong et al. (2011) applied the MCMC for stochastic image de-noising and found that their technique achieved excellent results in terms of both peak signal-to-noise ratio and mean structural similarity metrics when compared to other published approaches.

Nichols et al. (2011) applied the Markov Chain Monte Carlo technique for identifying cracks in a plate and their results show that this technique can approximate the state of damage in a cracked plate structure.

Deutch et al. (2011) successfully applied MCMC to play Trivia while Vrugt (2011) applied an adaptive Markov chain Monte Carlo simulation algorithm to solve discrete, non-continuous, posterior parameter estimation problems.

Wu and Drummond (2011) applied the MCMC technique for joint inference of microsatellite mutation models, population history and genealogies while Wöhling and Vrugt (2011) applied the MCMC for multi-response multi-layer vadose zone model calibration.

Jing and Vadakkepat (2009) applied a Markov Chain Monte Carlo technique to the tracking of maneuvering objects whereas Gallagher et al. (2009) applied the Markov Chain Monte Carlo procedure to identify optimal models, model resolution, and model selection for earth science problems. Curran (2008) applied the MCMC technique in DNA profiling. Other successful applications of the Markov Chain Monte Carlo technique include its use in environmental modeling (Gauchere et al. 2008), in medical imaging (Jun et al. 2008), in lake-water quality modeling (Malve et al. 2007), in economics (Jacquier et al. 2007), in statistics (Lombardi 2007), in decrypting classical cipher texts (Chen and Rosenthal 2011) and in robotics (Wang et al. 2011).

To apply the MCMC procedure, a system is considered whose evolution is characterized by a stochastic process consisting of random variables $\{x_1, x_2, x_3, \ldots, x_i\}$. A random variable x_i occupies a state x at discrete time i. The assembly of all possible states that all random variables can occupy is called a *state space*. If the probability that the system is in state x_{i+1} at time $i+1$ depends completely on the point that it was in state x_i at time i, then the random variables $\{x_1, x_2, x_3, \ldots, x_i\}$ form a Markov chain. In the Markov Chain Monte Carlo, the transition between states is achieved by adding random noise (ε) to the current state as follows (Bishop 1995; Marwala 2010; Marwala and Lagazio 2011):

$$x_{i+1} = x_i + \varepsilon \tag{4.5}$$

When the current state has been attained, it is either accepted or rejected. In this chapter the acceptance of a state is decided by applying the Metropolis algorithm (Bedard 2008; Meyer et al. 2008). This algorithm, developed by Metropolis et al. (1953) has been applied widely to solve problems in statistical mechanics. Bazavov et al. (2009) used biased Metropolis algorithms for protein simulation. Other applications of the Metropolis algorithms were in nuclear power plants (Sacco et al. 2008), in protein chains simulation (Tiana et al. 2007), and for the prediction of free Co-Pt nano-clusters (Moskovkin and Hou 2007). Restrepo-Parra et al. (2011) applied the Metropolis algorithm for the magnetic phase diagram simulation of $La_{1-x}Ca_xMnO_3$ system by using Metropolis algorithm while Beddard (2011) applied the Metropolis algorithm to calculate thermodynamic quantities in an undergraduate computational experiment. Santoso et al. (2011) applied a modified Metropolis-Hastings procedure with reduced chain correlation for efficient subset simulation for the reliability estimation of soil slope with spatially variable properties while Zuev and Katafygiotis (2011) applied a modified Metropolis-Hastings algorithm with delayed rejection to a subset simulation for computing small failure probabilities in high dimensions.

In conclusion, in the MCMC implementation, on sampling a stochastic process $\{x_1, x_2, x_3, \ldots, x_i\}$ comprised of random variables, random changes to x are presented by using Eq. 4.5 and they are either accepted or rejected, according to the following Metropolis et al. (1953) criterion (Marwala 2009; Marwala 2010; Marwala and Lagazio 2011):

$$\begin{aligned}
&if\, E_{new} < E_{old}\ accept\ state\ (s_{new})\\
&else\\
&accept\ (s_{new})\ with\ probability\\
&\exp\{-(E_{new} - E_{old})\}
\end{aligned} \tag{4.6}$$

Through a careful investigation of Eq. 4.6, it will be seen that states with high probability form the majority of the Markov chain, and those with low probability form the minority of the Markov chain.

4.3.3 Hybrid Monte Carlo

This chapter implements the Hybrid Monte Carlo (HMC) technique to estimate the posterior probability of the weight vectors, given the training data. This Monte Carlo method implements the gradient of the error that is calculated by applying a back-propagation technique. The usage of the gradient technique guarantees that the simulation does sample throughout the regions of higher probabilities and thus increases the time it takes to converge on a stationary probability distribution function. This method is regarded as a type of a Markov chain with transition between states attained by alternating between the 'stochastic' and 'dynamic moves'. The 'stochastic' moves permit the technique to explore states with different total energy whereas the 'dynamic' moves are achieved by applying the Hamiltonian dynamics and allowing the technique to search for states with the total energy nearly constant. In its basic form, the HMC technique can be regarded as a combination of Monte Carlo sampling technique which is steered by the gradient of the probability distribution function at each state.

Ghoufi and Maurin (2010) implemented the HMC technique to estimate the structural transitions of a porous Metal-organic framework material and confirmed that hybridizing the hybrid osmotic Monte Carlo technique with a "phase mixture" model is an effective technique to estimate the adsorption behavior accurately. Rei et al. (2010) implemented a hybrid Monte Carlo technique in a single vehicle routing problem with stochastic demands and their results showed that this technique is effective. Aleksandrov et al. (2010) implemented the HMC technique to study the vapor–liquid equilibria of copper and observed that the simulation and experiment were close. Zhang et al. (2010) implemented a hybrid Monte Carlo technique to simulate stress-induced texture evolution and used this result to construct an internal variable rate equation which could calculate the time evolution. Bogaerts (2009) implemented the HMC technique to study the effects of oxygen addition to argon glow discharges and Qian et al. (2011) implemented the hybrid Monte

Carlo technique to estimate the animal population affected by an environmental catastrophe. Kulak (2009) applied the HMC method to simulate fluorescence anisotropy decay whereas Suzuki et al. (2010) applied this technique in fluoride ion-water clusters.

Wendt et al. (2011) applied the hybrid Hybrid Monte Carlo technique in graphics while Hoefling et al. (2011) applied the hybrid Monte Carlo technique for modeling structural heterogeneity and quantitative Förster Resonance Energy Transfer efficiency distributions of polyprolines.

Cheng et al. (2011) applied hybrid Monte Carlo technique to study spacecraft thermal models and the results proved that it was superior to conventional approaches and fulfilled the necessities for thermal model correction while Zhang et al. (2011) applied the hybrid Monte Carlo method in stress-induced texture evolution with inelastic effects to develop a macroscopic equation that predicted such texture evolution.

Other applications of the hybrid Monte Carlo technique include its use in modeling probability distributions in Riemannian space (Paquet and Viktor 2011) and to estimate an animal population affected by an environmental catastrophe (Qian et al. 2011).

In statistical mechanics, the positions and the momentum of all molecules at a given time in a physical system is referred to as the *state space* of the system. The positions of the molecules describe the potential energy of the system and the momentum expresses the kinetic energy of the system. In this chapter, what is referred to in statistical mechanics as the *canonical distribution* of the 'potential energy' is the posterior distribution. The canonical distribution of the system's kinetic energy is (Neal 1993; Bishop 1995; Marwala 2009; Marwala and Lagazio 2011):

$$\begin{aligned} P(\{p\}) &= \frac{1}{Z_K} \exp(-K(\{p\})) \\ &= (2\pi)^{-n/2} \exp\left(-\frac{1}{2}\sum_i p_i^2\right) \end{aligned} \tag{4.7}$$

In molecular dynamics, p_i is the momentum of the ith molecule. At this juncture, p is not to be confused with, P, which stipulates the probability. In neural networks, p_i is a fictional parameter that is used to offer the method with molecular dynamics characteristics. It must be noted that the weight vector, $\{w\}$, and momentum vector, $\{p\}$, are of the same dimension and for that reason the superscript W is used in Eq. 4.3. The sum of the kinetic and potential energy is known as the *Hamiltonian* of the system and can be mathematically designated as follows (Neal 1993; Bishop 1995; Marwala 2009; Marwala and Lagazio 2011):

$$H(w, p) = \beta \sum^{N} \sum_{k}^{K} \{y_{nk} - t_{nk}\}^2 + \frac{\alpha}{2} \sum_{j=1}^{W} w_j^2 + \frac{1}{2} \sum_{i}^{W} p_i^2 \tag{4.8}$$

In Eq. 4.8, the first two expressions are the potential energy of the system, which is the exponent of the posterior distribution, and the last term is the kinetic energy. The canonical distribution over the phase space, *i.e.*, position and momentum, can be mathematically designated as follows (Neal 1993; Bishop 1995; Marwala 2009; Marwala and Lagazio 2011):

$$P(w, p) = \frac{1}{Z}\exp(-H(w, p)) = P(w|D)P(p) \tag{4.9}$$

By sampling through the weight and momentum space, the posterior distribution of weight is achieved by overlooking the distribution of the momentum vector, p. The dynamics in the phase space may be stated in terms of the Hamiltonian dynamics by articulating the derivative of the 'position' and 'momentum' in terms of fictional time τ. It should be recalled that the expression 'position' applied here is identical to the network weights. The dynamics of the system may thus be expressed through applying the Hamiltonian dynamics as follows (Neal 1993; Bishop 1995; Marwala 2009; Marwala and Lagazio 2010):

$$\frac{dw_i}{d\tau} = +\frac{\partial H}{\partial p_i} = p_i \tag{4.10}$$

$$\frac{dp_i}{d\tau} = +\frac{\partial H}{\partial w_i} = -\frac{\partial E}{\partial p_i} \tag{4.11}$$

The dynamics, stated in Eqs. 4.10 and 4.11, cannot be achieved exactly. As a result these equations are discretized by applying a 'leapfrog' technique. The leapfrog discretization of Eqs. 4.10 and 4.11 may be defined as follows (Neal 1993; Bishop 1995; Marwala 2009; Marwala and Lagazio 2011):

$$\hat{p}_i\left(\tau + \frac{\varepsilon}{2}\right) = \hat{p}_i(\tau) - \frac{\varepsilon}{2}\frac{\partial E}{\partial w_i}(\hat{w}(\tau)) \tag{4.12}$$

$$\hat{w}_i(\tau + \varepsilon) = \hat{w}_i(\tau) + \varepsilon\hat{p}_i\left(\tau + \frac{\varepsilon}{2}\right) \tag{4.13}$$

$$\hat{p}_i(\tau + \varepsilon) = \hat{p}_i\left(\tau + \frac{\varepsilon}{2}\right) - \frac{\varepsilon}{2}\frac{\partial E}{\partial w_i}(\hat{w}(\tau + \varepsilon)) \tag{4.14}$$

By applying Eq. 4.12, the leapfrog takes a slight half step for the momentum vector, $\{p\}$, and, applying Eq. 4.13, takes a full step for the 'position', $\{w\}$, and, by applying Eq. 4.14, takes a half step for the momentum vector, $\{p\}$. The combination of these three steps produce a single leapfrog iteration that calculates the 'position' and 'momentum' of a system at time $\tau + \varepsilon$ from the network weight vector and 'momentum' at time τ. The above discretization is reversible in time. It almost conserves the Hamiltonian, representing the total energy, and preserves the volume in the phase space, as required by Liouville's theorem (Neal 1993). The volume preservation is achieved since the moves that the leapfrog takes are shear transformations.

One subject that should be taken into account is that following Hamiltonian dynamics do not sample through the canonical distribution ergodically because the total energy stays the same, but at most samples through the micro-canonical distribution for a given energy. One technique applied to assure that the simulation is ergodic, is by applying 'stochastic' moves by changing the Hamiltonian, H, through the simulation and this is achieved by substituting the 'momentum' vector, $\{p\}$, before the next leapfrog iteration is attained. In this chapter, a normally distributed vector with a zero-mean alternates for the 'momentum' vector. The dynamic steps described in this section apply the gradient of the error with respect to the 'position', which is the network weight vector. The technique used to move from one state to another called the hybrid Monte Carlo which applies Hamiltonian dynamics to achieve dynamic moves and randomly changes the 'momentum' vector to attain stochastic moves. Simulating a distribution by perturbing a single vector, $\{w\}$ as is done in the MCMC is not practical because of the high dimensional nature of the state space and the variation of the posterior probability of the weight vector. A method that applies the gradient of the Hamiltonian with respect to the weight vector, $\{w\}$, was implemented to improve the Metropolis algorithm.

The Hybrid Monte Carlo method combines the stochastic dynamics model with the Metropolis algorithm, and in so doing eliminates the bias resulting from the use of a non-zero step size. The HMC technique operates by taking a series of trajectories from an initial state, *i.e.*, 'positions' and 'momentum', and moving in some direction in the state space for a given length of time and accepting the final state by applying the Metropolis algorithm. The validity of the hybrid Monte Carlo rests on three properties of the Hamiltonian dynamics. These properties have been described by Neal (1993), Bishop (1995), Marwala (2009) as well as Marwala and Lagazio (2011) as follows:

- Time reversibility: it is invariant under t→-t, p→-p.
- Conservation of energy: the $H(w,p)$ is the same at all times.
- Conservation of state space volumes due to Liouville's theorem (Neal 1993).

For a given leapfrog step size, ε_0, and the number of leapfrog steps, L, the dynamic transition of the hybrid Monte Carlo procedure is conducted as described by Neal (1993), Bishop (1995), Marwala (2009) as well as Marwala and Lagazio (2011):

1. Randomly choose the direction of the trajectory, λ, to be either -1 for a backwards trajectory or $+1$ for forwards trajectory.
2. Starting from the initial state, $(\{w\}, \{p\})$, perform L leapfrog steps with the step size $\varepsilon = \varepsilon_0(1 + 0.1k)$ resulting in state $(\{w\}^*, \{p\}^*)$. Here ε_0 is a selected fixed step size and k is a number selected from a uniform distribution and is between 0 and 1.

3. Reject or accept $(\{w\}^*, \{p\}^*)$ by applying the Metropolis criterion. If the state is accepted then the new state becomes $(\{w\}^*, \{p\}^*)$. If rejected the old state, $(\{w\}, \{p\})$, is retained as the new state.

After applying Step 3, the momentum vector is initiated before moving on to generate the following state. In this chapter, the momentum vector was sampled from a Gaussian distribution before producing the subsequent state. This ensures that the stochastic dynamics model samples are not limited to the micro-canonical ensemble. By altering the momentums, the total energy is allowed to change because the momentums of the particles are restored.

A note about the HMC method is that it applies the gradient information in Step 2 above by applying the leapfrog steps. The advantages of using this gradient information is that the HMC trajectories move in the direction of high probabilities, resulting in an improved probability that the resulting state be accepted and that the accepted states are not highly correlated. In neural networks the gradient is calculated using back-propagation (Bishop 1995).

The number of leapfrog steps, L, must be significantly higher than one to permit a fast exploration of the state space. The selection of ε_0 and L affects the speed at which the simulation converges to a stationary distribution and the correlation between the states accepted. The leapfrog discretization does not introduce systematic errors due to occasional rejection of states that result with the increase of the Hamiltonian. In Step 2 of the application of the HMC method, the step size $\varepsilon = \varepsilon_0(1 + 0.1k)$ where k is uniformly distributed between 0 and 1 is not fixed. In effect, this ensures that the definite step size for each trajectory is changed so that the accepted states do not have a high correlation. The same effect can be achieved by changing the leapfrog steps. In this chapter only the step size was changed. The application of the Bayesian approach to neural networks results in weight vectors that have a certain mean and standard deviation. As a result, the output parameters have a probability distribution. Following the rules of probability theory, the distribution of the output vector $\{y\}$ for a given input vector $\{x\}$ may be written in the following form as explained in Bishop (1995), Marwala (2009) and Marwala and Lagazio (2011):

$$p(\{y\}\left|\{x\}, D) = \int p(\{y\}\,|\{x\}, \{w\})\, p(\{w\}\,|D) d\{w\} \right. \tag{4.15}$$

In this chapter, the hybrid Monte Carlo method was implemented to determine the distribution of the weight vectors, and afterwards, of the output parameters. The integral in Eq. 4.15 may be estimated as follows (Bishop 1995; Neal 1993; Marwala 2009; Marwala and Lagazio 2011):

$$I \equiv \frac{1}{L}\sum_{i=1}^{L} f(\{w\}_i) \tag{4.16}$$

In Eq. 4.16, L is the number of retained states and f is the MLP network. An application of a Bayesian framework to the neural network results, with the mapping weight vector between the input and output having a probability distribution.

4.4 Fault Identification of Cylindrical Shells

In Chap. 3 the maximum-likelihood method was used to train the neural networks which were applied to identify faults in a population of cylindrical shells. In the same way, the Bayesian-formulated networks which are trained using the HMC were applied to classify faults in cylindrical shells. The networks were trained and tested using the data described in Chap. 3 and the architectures of the MLP networks were the same as those in Chap. 3. On implementing the hybrid Monte Carlo method to train the neural networks the following parameters were used (Marwala 2001):

- The number of initial states discarded, K, in the hope of reaching a stationary distribution was set to 100.
- The number of steps in each hybrid Monte Carlo trajectory was set to 100;
- The fixed step size was set to 0.001
- The number of samples retained to form a distribution was set to 500

The number of inputs and hidden units used are the same as those that were used in Chap. 3. The number of output units corresponds to the number of substructures. The data contribution to the error function was chosen arbitrarily. The fixed step size was selected through trial and error by examining how a selected step size influenced the acceptance rate of the states visited. This step size should be as close to zero as possible and if the step size is too low, then the dynamics of the hybrid Monte Carlo technique through the state space takes a long time to converge to a stationary posterior distribution; while if it is too large, then the process can possibly miss the stationary distribution.

The ability of the networks to detect the presence of faults and classify fault cases from the test data set is studied. When the trained networks are used to detect the presence of fault in the test data set as described in Chap. 3, the results in Tables 4.1 and 4.2 are obtained.

The confusion matrix attained when the pseudo-modal-energy-network was used is shown in Table 4.1. In this table 94.9% of [000]; all the one- and two-fault-cases; and 82.1% of [111] cases were correctly classified. The confusion matrix obtained when the modal-property-network was used is shown in Table 4.2. This table shows that this network classified 97.4% [000] fault-cases correctly; all the one- and two-fault-cases; and 66.7% of [111] fault-cases were classified correctly. These results show that the pseudo-modal-energy-network classifies fault-cases more accurately than the modal-property-network.

Table 4.1 Confusion matrix from the classification of fault cases in the test data using the pseudo-modal-energy-network

		Predicted							
		[000]	[100]	[010]	[001]	[110]	[101]	[011]	[111]
Actual	[000]	37	2	0	0	0	0	0	0
	[100]	0	3	0	0	0	0	0	0
	[010]	0	0	3	0	0	0	0	0
	[001]	0	0	0	3	0	0	0	0
	[110]	0	0	0	0	3	0	0	0
	[101]	0	0	0	0	0	3	0	0
	[011]	0	0	0	0	0	0	3	0
	[111]	0	0	0	0	5	1	1	32

Table 4.2 Confusion matrix from the classification of fault cases in the test data using the modal-energy-network

		Predicted							
		[000]	[100]	[010]	[001]	[110]	[101]	[011]	[111]
Actual	[000]	38	0	0	1	0	0	0	0
	[100]	0	3	0	0	0	0	0	0
	[010]	0	0	3	0	0	0	0	0
	[001]	0	0	0	3	0	0	0	0
	[110]	0	0	0	0	3	0	0	0
	[101]	0	0	0	0	0	3	0	0
	[011]	0	0	0	0	0	0	3	0
	[111]	0	0	0	0	5	2	6	26

4.5 Conclusion

Two Bayesian formulated neural networks were trained using pseudo-modal energies and modal properties were successfully used to perform fault identification in a population of cylindrical shells. The Bayesian networks were identified using the hybrid Monte Carlo technique. On average, it is found that the pseudo modal energies detect and classify faults more reliably than the modal properties do.

References

Akhmatskaya E, Bou-Rabee N, Reich S (2009) A comparison of generalized hybrid Monte Carlo methods with and without momentum flip. J Comput Phys 228:2256–2265

Aleksandrov T, Desgranges C, Delhommelle J (2010) Vapor–liquid equilibria of copper using hybrid Monte Carlo Wang-Landau simulations. Fluid Phase Equilibria 287:79–83

Bazavov A, Berg BA, Zhou H (2009) Application of biased metropolis algorithms: from protons to proteins. Math Comput Simul. doi:10.1016/j.matcom.2009.05.005

Bédard M (2008) Optimal acceptance rates for metropolis algorithms: moving beyond 0.234. Stoch Process Appl 118:2198–2222

Beddard GS (2011) Using the metropolis algorithm to calculate thermodynamic quantities: an undergraduate computational experiment. J Chem Educ 88(5):574–580
Bishop CM (1995) Neural networks for pattern recognition. Oxford University Press, Oxford
Bogaerts A (2009) Effects of oxygen addition to argon glow discharges: a hybrid Monte Carlo-fluid modeling investigation. Spectrochim Acta B: Atomic Spectrosc 64:1266–1279
Chen J, Rosenthal JS (2011) Decrypting classical cipher text using Markov Chain Monte Carlo. Stat Comput 22:397–413
Cheng WL, Liu N, Li Z, Zhong Q, Wang AM, Zhang ZM, He ZB (2011) Application study of a correction method for a spacecraft thermal model with a Monte-Carlo hybrid algorithm. Chin Sci Bull 56(13):1407–1412
Curran JM (2008) A MCMC method for resolving two person mixtures. Sci Justice 48:168–177
Daimon T (2008) Predictive checking for Bayesian interim analyses in clinical trials. Contemp Clin Trials 29(5):740–750. doi:10.1016/j.cct.2008.05.005, ISSN 1551–7144
Deutch D, Greenshpan O, Kostenko B, Milo T (2011) Using Markov Chain Monte Carlo to play Trivia. In: Proceedings of international conference on data engineering, art. no. 5767941, Hannover, Germany, pp 1308–1311
Do H, Wheatley RJ, Hirst JD (2010) Microscopic structure of liquid 1-1-1-2-tetrafluoroethane (R134a) from Monte Carlo simulation. Phys Chem Chem Phys 12:13266–13272
Droguett EL, das Chagas Moura M, Jacinto CM, Silva MF Jr (2008) A semi-Markov model with Bayesian belief network based human error probability for availability assessment of downhole optical monitoring systems. Simul Model Pract Theory 16(10):1713–1727, The Analysis of Complex Systems
Drugan MM, Thierens D (2010) Recombination operators and selection strategies for evolutionary Markov Chain Monte Carlo algorithms. Evol Intell 3:79–101
Dunn WL, Shultis JK (2009) Monte Carlo methods for design and analysis of radiation detectors. Radiat Phys Chem 78:852–858
Fefelov VF, Gorbunov VA, Myshlyavtsev AV, Myshlyavtseva MD (2009) The simplest self-assembled monolayer model with different orientations of complex organic molecules – Monte Carlo and transfer-matrix techniques. Chem Eng J 154:107–114
Feng Y, Schlindwein FS (2009) Normalized wavelet packets quantifiers for condition monitoring. Mech Syst Signal Process 23(3):712–723
Gallagher K, Charvin K, Nielsen S, Sambridge M, Stephenson J (2009) Markov Chain Monte Carlo (MCMC) sampling methods to determine optimal models, model resolution and model choice for earth science problems. Mar Pet Geol 26:525–535
Gauchere C, Campillo F, Misson L, Guiot J, Boreux JJ (2008) Parameterization of a process-based tree-growth model: comparison of optimization. MCMC and particle filtering algorithms. Environ Model Software 23:1280–1288
Ghoufi A, Maurin G (2010) Hybrid Monte Carlo simulations combined with a phase mixture model to predict the structural transitions of a porous metal-organic framework material upon adsorption of guest molecules. J Phys Chem C 114:6496–6502
Hoefling M, Lima N, Haenni D, Seidel CAM, Schuler B, Grubmuller H (2011) Structural heterogeneity and quantitative FRET efficiency distributions of polyprolines through a hybrid atomistic simulation and Monte Carlo approach. PLoS One 6(5):e19791
Hu J, Zhang L, Ma L, Liang W (2010) An integrated method for safety pre-warning of complex system. Saf Sci 48(5):580–597
Jacquier E, Johannes M, Polson N (2007) MCMC maximum likelihood for latent state models. J Econometrics 137:615–640
Jia Y, Zhang C (2009) Front-view vehicle detection by Markov Chain Monte Carlo method. Pattern Recognit 42:313–321
Jing L, Vadakkepat P (2009) Interacting MCMC particle filter for tracking maneuvering target. Digit Signal Process. doi:10.1016/j.dsp. 2009.08.011
Jun SC, George JS, Kim W, Pare-Blagoev J, Plis S, Ranken DM, Schmidt DM (2008) Bayesian brain source imaging based on combined MEG/EEG and fMRI using MCMC. Neuroimage 40:1581–1594

Kandela B, Sheorey U, Banerjee A, Bellare J (2010) Study of tablet-coating parameters for a pan coater through video imaging and Monte Carlo simulation. Powder Technol 204:103–112

Katsis CD, Goletsis Y, Rigas G, Fotiadis DI (2011) A wearable system for the affective monitoring of car racing drivers during simulated conditions. Transportation Res C: Emerg Technol 19(3):541–551

Kim MJ, Jiang R, Makis V, Lee C-G (2011) Optimal Bayesian fault prediction scheme for a partially observable system subject to random failure. Eur J Oper Res 214:331–339

Kohda T, Cui W (2007) Risk-based reconfiguration of safety monitoring system using dynamic Bayesian network. Reliability Eng Syst Saf 92(12):1716–1723, Special Issue on ESREL 2005

Kulak L (2009) Hybrid Monte-Carlo simulations of fluorescence anisotropy decay in three-component donor-mediator-acceptor systems in the presence of energy transfer. Chem Phys Lett 467:435–438

Lagazio M, Marwala T (2005) Assessing different Bayesian neural network models for militarized interstate dispute. Soc Sci Comput Rev 2005:1–12

Lai Y (2009) Adaptive Monte Carlo methods for matrix equations with applications. J Comput Appl Math 231:705–714

Liesenfeld R, Richard J (2008) Improving MCMC, using efficient importance sampling. Comput Stat Data Anal 53:272–288

Liu X, Newsome D, Coppens M (2009) Dynamic Monte Carlo simulations of binary self-diffusion in ZSM-5. Microporous Mesoporous Mater 125:149–159

Lombardi MJ (2007) Bayesian inference for [alpha]-stable distributions: a random walk MCMC approach. Comput Stat Data Anal 51:2688–2700

MacKay DJC (1991) Bayesian methods for adaptive models. PhD thesis, California Institute of Technology

MacKay DJC (1992) A practical Bayesian framework for backpropagation networks. Neural Comput 4:448–472

Malve O, Laine M, Haario H, Kirkkala T, Sarvala J (2007) Bayesian modelling of algal mass occurrences – using adaptive MCMC methods with a lake water quality model. Environ Model Software 22:966–977

Martin C, Ayesa E (2010) An Integrated Monte Carlo Methodology for the calibration of water quality models. Ecol Model 221:2656–2667

Marwala T (2001) Fault identification using neural networks and vibration data. PhD thesis, University of Cambridge

Marwala T (2007) Bayesian training of neural network using genetic programming. Pattern Recognit Lett. doi:org/10.1016/j.patrec.2007.034

Marwala T (2009) Computational intelligence for missing data imputation, estimation and management: knowledge optimization techniques. IGI Global Publications, New York

Marwala T (2010) Finite element model updating using computational intelligence techniques. Springer, London

Marwala T, Lagazio M (2010) Militarized conflict modeling using computational intelligence. Springer, London

Marwala T, Lagazio M (2011) Militarized conflict modeling using computational intelligence techniques. Springer, London

Mathe P, Novak E (2007) Simple Monte Carlo and the metropolis algorithm. J Complex 23:673–696

McClarren RG, Urbatsch TJ (2009) A modified implicit Monte Carlo method for time-dependent radiative transfer with adaptive material coupling. J Comput Phys 228:5669–5686

Metropolis N, Rosenbluth A, Rosenbluth M, Teller A, Telle E (1953) Equation of state calculations by fast computing machines. J Chem Phys 21:1087–1092

Meyer R, Cai B, Perron F (2008) Adaptive rejection metropolis sampling using Lagrange interpolation polynomials of degree 2. Comput Stat Data Anal 52:3408–3423

Moller M (1993) A scaled conjugate gradient algorithm for fast supervised learning. Neural Netw 6:525–533

Moreau M, Laffly D, Joly D, Brossard T (2005) Analysis of plant colonization on an arctic moraine since the end of the Little Ice Age using remotely sensed data and a Bayesian approach. Remote Sens Environ 99(3):244–253. doi:10.1016/j.rse.2005.03.017, ISSN 0034–4257

Moskovkin P, Hou M (2007) Metropolis Monte Carlo predictions of free Co-Pt nanoclusters. J Alloys Compounds 434–435:550–554

Neal RM (1993) Probabilistic inference using Markov Chain Monte Carlo methods. University of Toronto technical report CRG-TR-93-1, Toronto, Canada

Nebot JVA, Morales-Menendez R, Guevara AJV, Rodriguez CA (2007) Surface roughness and cutting tool-wear diagnosis based on Bayesian networks. In: Zhang H-Y (ed) Fault detection, supervision and safety of technical processes 2006. Elsevier, Oxford, pp 408–413

Nichols JM, Moore EZ, Murphy KD (2011) Bayesian identification of a cracked plate using a population-based Markov Chain Monte Carlo method. Comput Struct 89(13–14):1323–1332

Oliveira RG, Schneck E, Quinn BE, Konovalov OV, Brandenburg K, Seydel U, Gill T, Hanna CB, Pink DA, Tanaka M (2009) Physical mechanisms of bacterial survival revealed by combined grazing-incidence X-ray scattering and Monte Carlo simulation. Comptes Rendus Chimie 12:209–217

Ozaki M, Ohno M, Terada Y, Watanabe S, Mizuno T, Takahashi T, Kokubun M, Tsujimoto M, Yamasaki NY, Odaka H, Takei Y, Yuasa T, Furuzawa A, Mori H, Matsumoto H, Okajima T, Kilbourne CA, Tajima H, Ishisaki Y (2010) The Monte Carlo simulation framework of the ASTRO-H X-Ray Observatory. In: Proceedings of SPIE – the international society for optical engineering:7732, San Diego, California, art. no. 773239

Padilla Cabal F, Lopez-Pino N, Luis Bernal-Castillo J, Martinez-Palenzuela Y, Aguilar-Mena J, D'Alessandro K, Arbelo Y, Corrales Y, Diaz O (2010) Monte Carlo based geometrical model for efficiency calculation of an N-type HPGe detector. Appl Radiat Isot 68:2403–2408

Paquet E, Viktor HL (2011) Probability distributions from Riemannian geometry, generalized hybrid Monte Carlo sampling and path integrals. In: Proceedings of SPIE – the International Society for Optical Engineering, 7864, San Francisco, California, art. no. 78640X

Pollino CA, White AK, Hart BT (2007) Examination of conflicts and improved strategies for the management of an endangered Eucalypt species using Bayesian networks. Ecol Model 201(1): 37–59, Management, control and decision making for ecological systems

Qi F, Huang B (2011) Bayesian methods for control loop diagnosis in the presence of temporal dependent evidences. Automatica 47(7):1349–1356

Qian G, Li N, Huggins R (2011) Using capture-recapture data and hybrid Monte Carlo sampling to estimate an animal population affected by an environmental catastrophe. Stat Data Anal 55(1):655–666

Rahmati M, Modarress H (2009) Nitrogen adsorption on nanoporous zeolites studied by grand canonical Monte Carlo simulation. J Mol Struct: Theochem 901:110–116

Ratick S, Schwarz G (2009) Monte Carlo simulation. In: Kitchin R, Thrift N (eds) International encyclopedia of human geography. Elsevier, Oxford

Rei W, Gendreau M, Soriano P (2010) A hybrid Monte Carlo local branching algorithm for the single vehicle routing problem with stochastic demands. Trans Sci 44:136–146

Restrepo-Parra E, Salazar-Enrquez CD, Londoo-Navarro J, Jurado JF, Restrepo J (2011) Magnetic phase diagram simulation of La1-xCaxMnO 3 system by using Monte Carlo, Metropolis algorithm and Heisenberg model. J Magn Magn Mater 323(11):1477–1483

Robert CP, Casella G (2004) Monte Carlo statistical methods. Springer, London

Rodina A, Bliznakova K, Pallikarakis N (2010) End stage renal disease patients' projections using Markov Chain Monte Carlo simulation. Proc IFMBE 29:796–799

Roskilly SJ, Colbourn EA, Alli O, Williams D, Paul KA, Welfare EH, Trusty PA (2010) Investigating the effect of shape on particle segregation using a Monte Carlo simulation. Powder Technol 203:211–222

Sacco WF, Lapa CMF, Pereira CMNA, Filho HA (2008) A metropolis algorithm applied to a nuclear power plant auxiliary feedwater system surveillance tests policy optimization. Prog Nucl Energy 50:15–21

Santoso AM, Phoon KK, Quek ST (2011) Modified Metropolis-Hastings algorithm with reduced chain correlation for efficient subset simulation. Probabilistic Eng Mech 26(2):331–341

Subrahmanya N, Shin YC, Meckl PH (2010) A Bayesian machine learning method for sensor selection and fusion with application to on-board fault diagnostics. Mech Syst Signal Proces 24(1):182–192

Suzuki K, Tachikawa M, Shiga M (2010) Efficient ab initio path integral hybrid Monte Carlo based on the fourth-order Trotter expansion: application to fluoride ion-water cluster. J Chem Phys 132: art. no. 144108. doi:10.1063/1.3367724

Tiana G, Sutto L, Broglia RA (2007) Use of the metropolis algorithm to simulate the dynamics of protein chains. Phys A: Stat Mech Appl 380:241–249

Vrugt JA (2011) DREAM(D): an adaptive Markov Chain Monte Carlo simulation algorithm to solve discrete, noncontinuous, posterior parameter estimation problems. Hydrol Earth Syst Sci Discuss 8(2):4025–4052

Wang H, Harrison KW (2010) Adaptive Bayesian contaminant source characterization in water distribution systems via a parallel implementation of Markov Chain Monte Carlo (MCMC). In: Proceedings of the World Environmental and Water Resources Congress, Providence, Rhode Island, pp 4323–4329

Wang XY, Wong A, Ho P-H (2010) Spectrum sensing in cognitive radio using a Markov-Chain Monte-Carlo scheme. IEEE Commun Lett 14:830–832

Wang Y, Wu H, Handroos H (2011) Markov Chain Monte Carlo (MCMC) methods for parameter estimation of a novel hybrid redundant robot. Fusion Eng Des 86:1863–1867

Weidl G, Madsen AL, Israelson S (2005) Applications of object-oriented Bayesian networks for condition monitoring, root cause analysis and decision support on operation of complex continuous processes. Comput Chem Eng 29(9):1996–2009

Wendt KA, Drut JE, Lahde TA (2011) Toward large-scale hybrid Monte Carlo simulations of the Hubbard model on graphics processing units. Comput Phys Commun 182(8):1651–1656

Willis AJ (2010) Condition monitoring of centrifuge vibrations using kernel PLS. Comput Chem Eng 34(3):349–353

Wöhling T, Vrugt JA (2011) Multiresponse multilayer vadose zone model calibration using Markov Chain Monte Carlo simulation and field water retention data. Water Resour Res 47(4), art. no. W04510. doi:10.1029/2010WR009265

Wong A, Mishra A, Zhang W, Fieguth P, Clausi DA (2011) Stochastic image denoising based on Markov-Chain Monte Carlo sampling. Signal Process 91(8):2112–2120

Wu CH, Drummond AJ (2011) Joint inference of microsatellite mutation models, population history and genealogies using transdimensional Markov Chain Monte Carlo. Genetics 188: 151–164

Xia J, Liu L, Xue J, Wang Y, Wu L (2009) Modeling of radiation-induced bystander effect using Monte Carlo methods. Nucl Instrum Methods Phys Res Sect B: Beam Interact Mater Atoms 267:1015–1018

Yu AJ, Dayan P, Cohen JD (2009) Dynamics of attentional selection under conflict: toward a rational Bayesian account. J Exp Psychol Hum Percept Perform 35(3):700–717. doi:10.1037/a0013553, ISSN 0096–1523

Yuen K-V, Kuok S-C (2010) Ambient interference in long-term monitoring of buildings. Eng Struct 32(8):2379–2386

Zhang L, Bartel T, Lusk MT (2010) Parallelized hybrid Monte Carlo simulation of stress-induced texture evolution. Comput Mater Sci 48:419–425

Zhang L, Dingreville R, Bartel T, Lusk MT (2011) Hybrid Monte Carlo simulation of stress-induced texture evolution with inelastic effects. Metallurgical Mater Trans A: Phys Metallurgy Mater Sci 42(3):575–581

Zhao H, Zheng C (2009) Correcting the multi-Monte Carlo method for particle coagulation. Powder Technol 193:120–123

Zuev KM, Katafygiotis LS (2011) Modified Metropolis-Hastings algorithm with delayed rejection. Probabilistic Eng Mech 26(3):405–412

Chapter 5
The Committee of Networks Approach to Condition Monitoring

5.1 Introduction

The identification of faults in mechanical systems at the manufacturing stage offers considerable economic benefits. By identification, this book means that the fault is as (1) detected; (2) located; and (3) the extent of the fault is quantified. Vibration methods are some of the many techniques that have been implemented with varying degrees of success to identify mechanical faults (Friswell and Mottershead 1995; Doebling et al. 1996; Marwala 2000). It has been shown that the success of fault identification methods depends on the type of signal implemented for diagnostics (Marwala and Heyns 1998; Marwala 2000). There are three types of signals that may be implemented for fault identification purposes and, as discussed in Chap. 2, these are the modal properties (mode shapes, as well as damping and natural frequencies), the Frequency Response Functions (FRFs), and the Wavelet Transform (WT) data. Modal properties are the easiest to implement, but are most suitable for detecting large faults. The modal properties approaches are not successful when used to identify faults in structures that are highly damped. They also necessitate measurements at many points and do not work well for nonlinear structural diagnosis. One limitation of FRF data is that they contain a great deal of extra information between resonance peaks. It is not clear how best to select the frequency bandwidth of interest. Also, FRF data are normally noisy at the anti-resonance regions, and there is no technique for choosing how to process FRFs for a specific problem. Nevertheless, FRF techniques have the following benefits (Imregun et al. 1995; Marwala 2000):

- Measurements at many points on the structure are not necessarily required.
- Estimated modal properties are further from measured data (in the time domain) than are FRFs. This is because modal properties are identified from FRFs using modal analysis. Using FRFs directly avoids errors incurred during modal analysis.

T. Marwala, *Condition Monitoring Using Computational Intelligence Methods: Applications in Mechanical and Electrical Systems*, DOI 10.1007/978-1-4471-2380-4_5,

- The FRF approach is applicable to non-modal behavior, for example, in cases of higher damping and modal density.
- It is possible to check a given solution by generating another one, since the problem is over determined due to the availability of FRFs at numerous excitation frequencies. This offers the possibility of using statistical techniques to determine confidence parameters and to interpret the results obtained.

The main limitation of the wavelet method is that there are many types of wavelets and there is no systematic technique to select the most suitable WT for fault identification. Wavelets generate excessive information and may be applied by monitoring a number of parameters. Wavelet techniques have the following advantages (Marwala 2000):

- WTs are relatively sensitive to local defects.
- Measurements at many points on the structure are not necessarily required.
- The WT approach is applicable to non-modal behavior, for example, in cases of higher damping and modal density.
- The problem is over-determined due to WT data at numerous frequencies and time domains.
- WTs are effective in the identification of damage that results in the loss of linearity of a structure.

For this chapter, three independent back-propagation (multi-layer perceptron) neural networks (Bishop 1995) were trained using modal properties, FRFs, and WT data. These were used in parallel to diagnose faults in a population of cylindrical shells. It was found that the committee technique was more reliable than using each method in isolation. The idea of using neural networks in parallel was conceived by Perrone and Cooper (1993). Levin and Lieven (1998) implemented modal properties in conjunction with neural networks to identify faults on a cantilevered beam. Atalla and Inman (1998) implemented FRFs to identify faults in finite-element models. Marwala and Hunt (1999) implemented modal properties and FRFs simultaneously to identify faults. Paya et al. (1997) implemented WT data and neural networks to identify faults on rotating machinery. The committee method is the subject of the next section.

5.2 A Committee of Networks

For this chapter a committee of networks is applied for classification of fault in a population of cylinders. Du et al. (2007) successfully implemented a committee of probabilistic radial basis function neural networks to identify palm prints, while Marwala et al. (2001) applied a committee of agents and genetic programming to evolve a stock market prediction system. Anthony (2007) studied the generalization error of fixed combinations of classifiers, while Sheikh-Ahmad et al. (2007) used the committee of neural network for force prediction models in a milling process.

Marwala (2001) applied a probabilistic fault identification process in structures using a committee of neural networks and vibration data.

Abdel-Aal (2005a) presented a three-member committee of multi-layer perceptron networks for improving electric load forecasts and observed that the committee decreases forecasting errors when compared with individual networks. Furthermore, the author extended the application of the committee of networks to the problem of modeling medical data, and found that the committee method offered a decrease in classification errors of up to 20% when compared to stand-alone networks. In the committee method, it is known that the existence of diversity within the members of the committee improves the performance of the committee. Abdel-Aal (2005b) presented diversity by training several members of the committee with different data.

Karimpouli et al. (2010) applied a committee of neural networks to predict the permeability of petroleum reservoirs, while Kadkhodaie-Ilkhchi et al. (2009) applied a committee of neural networks for the prediction of normalized oil content from logged oil-well data from South Pars Gas Field in the Persian Gulf.

Jafari et al. (2011) applied a committee of neural networks with a fuzzy genetic algorithm to predict a reservoir parameter in the petroleum industry, while van Hinsbergen et al. (2009) applied a Bayesian committee of neural networks to predict travel times.

Other successful implementations of the committee method that revealed improvement over individual methods include an application for human face recognition by Zhao et al. (2004), recognition of the swallow acceleration signal by Das et al. (2001), choosing salient features by Bacauskiene and Verikas (2004), speaker verification by Reddy and Buch (2003), automatic fire detection (Fernandes et al. 2004) as well as permeability prediction by Chen and Lin (2006), and missing data estimation (Marwala 2009).

The committee approach presented in this chapter falls within a family of techniques called ensembles of networks. There are many types of network ensembles, and these include the Bayes Optimal Classifier, Bayesian model averaging, bagging, boosting and stacking. Some of these techniques are described in the next section.

5.2.1 Bayes Optimal Classifier

The Bayes Optimal Classifier (BOC) is an optimal classification method which is an ensemble of all the hypotheses in the hypothesis space and, therefore, no other ensemble can perform better than it (Bishop 1995). The vote of each hypothesis is proportional to the likelihood that the training data set is sampled from a system where that hypothesis is true. To enable training data of finite size, each hypothesis' vote is also multiplied by the prior probability of that hypothesis. The BOC can then be mathematically written as follows (Bishop 1995):

$$y = \arg\max_{c_j \in C} \sum_{h_i \in H} P\left(c_j \middle| h_i\right) P\left(T \middle| h_i\right) P\left(h_i\right) \tag{5.1}$$

Here y is the estimated class, C is a full set with all classes, H is the hypothesis space, P is the probability, and T is the training data set. The BOC denotes a hypothesis that is not necessarily located in H. The hypothesis characterized by the BOC is the optimal hypothesis in the ensemble space; however, the BOC can only be applied to the simplest of problems. There are a number of explanations why the BOC cannot be practically applied, including the fact that the hypothesis spaces are too large to iterate over. Many hypotheses give only a predicted class instead of the probability for each class. Calculating the unbiased estimate of the probability of the training set, given a hypothesis, is difficult and computing the prior probability for each hypothesis is not feasible.

5.2.2 Bayesian Model Averaging

Bayesian model averaging is an ensemble method that aims to estimate the BOC by sampling hypotheses from the hypothesis space and combining these hypotheses using a Bayes' framework (Hoeting et al. 1999). As opposed to the Bayes optimal classifier described in the last section, Bayesian model averaging can be practically applied using procedures such as Monte Carlo sampling methods (Marwala 2009). It has been shown that, under certain conditions, when hypotheses are sampled and averaged according to Bayes' theorem, this procedure produces an expected error that is at most twice the expected error of the BOC (Haussler et al. 1994). Nevertheless, Bayesian model averaging has the shortcoming of over-fitting and performs worse empirically than simple ensembles do, for instance bagging (Domingos 2000).

Park and Grandhi (2011) successfully applied the Bayesian model averaging to combine the predictions of a system response into a single prediction and applied this to a nonlinear spring-mass system, a laser peening process, and a composite material.

Other successful applications of Bayesian model averaging include the estimation of the mortality risk associated with heat waves (Bobb et al. 2011), the forecasting of the monthly industrial production output of six countries (Feldkircher 2011), the assessment of environmental stressors (Boone et al. 2011), the analysis of schizophrenia family data (Tsai et al. 2011), quantifying multiple types of uncertainty in computer simulation (Park and Grandhi 2010), atmospheric dispersion studies (Potempski et al. 2010) and microarray data survival analysis (Bichindaritz and Annest 2010).

5.2.3 Bagging

Bagging, which is also called Bootstrap averaging, is a technique based on a combination of models fitted to bootstrap samples of a training data set to reduce the variance of the prediction model (Breiman 1996). Bagging essentially entails randomly choosing a part of the training data, using this part to train a model and then repeating this process. Thereafter, all trained models are combined with equal weights to form an ensemble.

Pino-Mejias et al. (2008) applied a reduced bootstrap aggregating learning algorithms to simulated classification and regression problems. They applied the reduced bootstrap for bagging unstable learning algorithms such as decision trees and neural networks and found that the technique reduced variance.

Kyung and Lee (1999) applied a bootstrap and aggregating classifier for speaker recognition. Experiments were donc on a closed set, text-independent and speaker identification system using the TIMIT database and their method demonstrated significantly improved performance over the conventional classifier.

Louzada et al. (2011) applied poly-bagging predictors in classification modeling of credit scoring. Their bagging technique consisted of combining predictors over a succession of re-sampling and their results showed that the poly-bagging method improved the modeling performance measures while retaining a flexible structure which was easy to apply.

Jia et al. (2011) generalized the selective clustering ensemble procedure and presented a selective spectral clustering ensemble where the component clustering of the ensemble system were generated by spectral clustering capable of engendering diverse committees. The random scaling parameter, a Nyström approximation was applied to perturb spectral clustering for creating the components of the ensemble system. Subsequent to the production of component clustering, the bagging method was applied to evaluate the component clustering. The results showed that the method achieved improved results over the conventional clustering ensemble methods.

Yu (2011) presented weighted bagging for a regularization technique and applied this to some real data sets, while Hernandez-Lobato et al. (2011) conducted an empirical analysis and evaluated approximate techniques for pruning regression bagging.

Other applications of bagging were for bi-clustering of gene expression data (Hanczar and Nadif 2011), while Hu et al. (2011) applied bagging in hybrid modeling for the prediction of leaching rate in a leaching process based on negative correlation learning, and Osawa et al. (2011) applied bagging in the analysis of zero-inflated data.

5.2.4 Boosting

Boosting is a technique that incrementally builds an ensemble by training each new model with data that the previously trained model mis-classified. Then the ensemble, which is a combination of all trained models, is used for prediction. Jasra and Holmes (2011) applied stochastic boosting algorithms which used sequential Monte Carlo methods. It was observed that stochastic boosting provided better predictions for classification problems than the conventional boosting algorithm. Leitenstorfer and Tutz (2011) applied boosting methods to estimate single-index models, while Baras et al. (2011) applied Bayesian networks for the automatic boosting of cross-product coverage. Furthermore, Kajdanowicz and Kazienko (2011) studied the structured output element ordering in boosting-based classification. Khoshgoftaar et al. (2011) compared boosting and bagging methods by applying these techniques to noisy and imbalanced data and observed that the bagging methods usually outperform boosting.

5.2.5 Stacking

The critical prior belief in the scientific method is that one can select from a set of models by comparing them on data that was not used to create the models. This prior belief can also be used to select amongst a set of models based on a single data set by using a technique called cross-validation (Bishop 1995). This is conducted by dividing the data set into a *held-in* data set, which is used to create the models, and a *held-out* data set which is used to test the created models (Sill et al. 2009).

Stacking takes advantage of this prior belief further by using performance on the held-out data to combine the models instead of selecting from them the best performing model when tested on the held-out data. This is done because the ensemble usually performs better than any single one of the trained models (Wolpert 1992). It has been successfully applied in both supervised learning (regression), unsupervised learning (density estimation) and to approximate Bagging's error rate (Breiman 1996; Smyth and Wolpert 1999; Wolpert and Macready 1999; Rokach 2010). The stacking method has been observed to perform better than the Bayesian model-averaging technique (Clarke 2003).

Drygajlo et al. (2011) applied a generalized stacking model for adult face recognition in score-age-quality classification space, while Larios et al. (2011) applied a stacking in an object-class recognition method to combine scores from random trees.

Shiraishi and Fukumizu (2011) presented a method for combining binary classifiers which trained a combining method of binary classifiers using statistical techniques such as penalized logistic regression, stacking, and a sparsity promoting penalty. Their method outperformed conventional classifiers and approximated conditional probability for each class.

Tang et al. (2010) applied re-ranking for stacking ensemble learning where the predictive scores of the base classifiers were assembled by the meta-learner and re-ranked according to the scores. Their method could find the best linear combination of the base classifiers on the training samples. When their method was tested on a number of public datasets, it was observed that the proposed algorithm outperformed the baseline procedures and several state-of-the-art stacking processes.

Homayouni et al. (2010) presented a Lazy Stacking (LS) method for building a classifier ensemble learner and tested this method against four rival procedures on a large suite of ten real-world benchmark numeric datasets. The results obtained confirmed that LS can outperform other approaches.

Other successful applications of stacking include remote sensing (Huang et al. 2011), credit scoring (Wang et al. 2011), and aging face verification (Li et al. 2010).

5.2.6 *Evolutionary Committees*

Evolutionary committees are techniques for making the construction of the committee method adapt in line with the environment. This is usually done through evolving the weighting function that defines the contribution of each individual method, with respect to the overall outcome of the committee.

Marwala (2009) introduced committees of networks for missing data estimation. The first committee of networks was made of Multi-Layer Perceptrons (MLPs), Support Vector Machines (SVMs), and Radial Basis Functions (RBFs); and entailed the weighted combination of these three networks. The second, third, and fourth committees of networks were evolved using a genetic programming method and used the MLPs, RBFs and SVMs, respectively. The committees of networks were applied, collectively, with a hybrid particle swarm optimization and genetic algorithm technique for missing data estimation. When they were tested on an artificial taster as well as HIV datasets and then compared to the individual MLPs, RBFs, and SVMs for missing data estimation, the committee of network approach was observed to give better results than the three approaches acting in isolation. Nonetheless, this improvement came at a higher computational load than the individual methods. In addition, it was observed that evolving a committee technique was a good way of constructing a committee.

Evolving networks has been the topic of study for some time (Marwala 2009). Rajan and Mohan (2007) applied an evolutionary programming method which was based on simulated annealing to solve the unit commitment problem, while Basu (2004) applied an evolutionary programming technique to create an interactive fuzzy satisfying scheme and applied this to solve a multi-objective short-term hydrothermal scheduling. Shi and Xu (2001) applied a self-adaptive evolutionary programming technique and used this to optimize the multi-objective operation of power systems and Cao et al. (2000) applied an evolutionary programming technique to a mixed-variable optimization problem.

5.3 Theoretical Background

The dynamics of any structure may be expressed in terms of mass, damping, and stiffness matrices as well as the acceleration, velocity, and displacement vector (Ewins 1995). The structure may be excited using an impulse hammer and the response can be measured by using an accelerometer. These responses may be transformed into other properties, such as FRFs, modal properties, and WT data (Newland 1993; Maia and Silva 1997; Marwala 2000). The mass, damping, and stiffness matrices depend on physical parameters such as the density, Poisson's ratio, and the Young's modulus of each structural member. Therefore, there are relationships between physical properties of the structure and the frequency response functions, modal properties, and wavelet transform (Marwala 1999, 2000). In the next section these relationships will be identified.

5.3.1 *Pseudo Modal Energies Method*

The Fast Fourier Transform (FFT) can be applied to both the excitation (i.e., acceleration) and the response to obtain the FRFs. The FRF is defined as the ratio of the transformed response to the transformed excitation (Ewins 1995). The pseudo-modal energies, denoted α, are integrals of the real and imaginary components of the frequency response functions over chosen frequency ranges that bracket the natural frequencies (Marwala 2001), as explained in Chap. 2. The pseudo-modal energy matrix is related to the spatial properties of the structure and if adequate, data that defines the relationship between changes in physical parameters and changes in pseudo modal energies may be produced. From this set of data, a functional mapping between the identity of fault y_1 and the pseudo modal energy vector $\boldsymbol{\alpha}$ may be represented in the following form:

$$y_1 = f(\boldsymbol{\alpha}) \tag{5.2}$$

5.3.2 *Modal Properties*

From the FRF data, modal properties which are natural frequencies and modal properties may be extracted using a process called modal analysis (Ewins 1995; Maia and Silva 1997). The modal properties data are related to the spatial properties of the structure and adequate data that defines the relationship between changes in physical parameters and changes in modal properties may be produced. Similarly, a functional mapping between the identity of fault y_2 and the modal properties vector χ may be quantified by the following equation:

$$y_2 = f(\chi) \tag{5.3}$$

5.3.3 *Wavelet Transforms (WT)*

The wavelet transform of a signal is an illustration of a time-frequency decomposition, which highlights the local features of a signal (Daubechie 1991) and was discussed in Chap. 2. The WT in this chapter is from the orthogonal wavelet family (Daubechie 1991) defined by Newland (1993). The relationship between the physical properties of the structure and the WT of the impulse of a unit magnitude may be used to identify faults on structures. Liew and Wang (1998) applied WT data to identify damage in structures. A functional mapping between the identity of fault y_3 and the WT of the response vector κ may be quantified by the following equation:

$$y_3 = f(\kappa) \tag{5.4}$$

5.3.4 *Neural Networks*

For this chapter, neural networks were viewed as parameterized graphs that make probabilistic assumptions about data. Learning algorithms were viewed as methods for finding parameter values that look probable in light of the data. Supervised learning is the case where the input and the output are available, and neural networks are used to approximate the functional mapping between the two. The type of neural network applied for this chapter was the multi-layer perceptron (Jordan and Bishop 1996). The multi-layer perceptron can approximate any continuous function to an arbitrary accuracy if the number of hidden units is sufficiently large. For this chapter, the output units represent the identity of faults and the inputs are the pseudo-modal energies, modal properties or wavelet transform.

5.4 Theory of Committee of Networks

For this chapter, a committee of networks technique, as illustrated in Fig. 5.1, was introduced. The committee method in this figure comprises three networks and the output is the weighted average of the outputs of these three networks. The ideas presented in this section are the adaptation and the addition of the work by Perrone and Cooper (1993) who introduced the concept of a committee of networks and it was extended and applied to mechanical systems by Marwala and Hunt (1999), as well as Marwala (2000). It is confirmed that a committee of networks provides results that are more reliable than when using networks in isolation for fault identification.

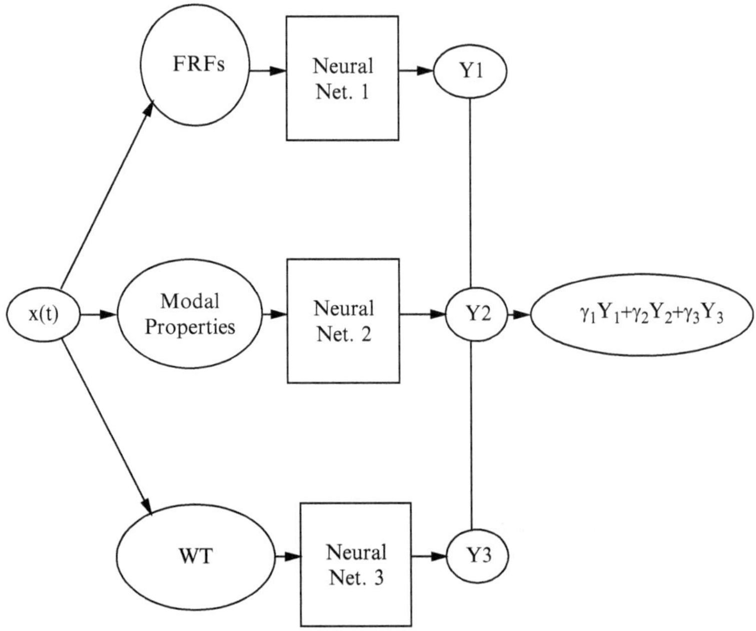

Fig. 5.1 Illustration of committee of networks

The mapping of the FRFs, modal properties, and wavelet data to the identities of faults (y_1, y_2, and y_3) may be written as the desired function plus an error. For notational accessibility, the mapping functions are assumed to have single outputs y_1, y_2, and y_3. This can be easily adapted to multiple outputs as follows (Perrone and Cooper 1993; Marwala 2000):

$$y_1(\boldsymbol{\alpha}) = h(\boldsymbol{\alpha}) + e_1(\boldsymbol{\alpha}) \tag{5.5}$$

$$y_2(\boldsymbol{\chi}) = h(\boldsymbol{\chi}) + e_2(\boldsymbol{\chi}) \tag{5.6}$$

$$y_3(\boldsymbol{\kappa}) = h(\boldsymbol{\kappa}) + e_3(\boldsymbol{\kappa}) \tag{5.7}$$

Here $h(\cdot)$ is approximated mapping function; and $e(\cdot)$ is the error. The mean square errors (MSE) for model $y_1(\boldsymbol{\alpha})$, $y_2(\boldsymbol{\chi})$, and $y_3(\boldsymbol{\kappa})$ may be written as follows (Perrone and Cooper 1993):

$$E_1 = \varepsilon\left[\{y_1(\boldsymbol{\alpha}) - h(\boldsymbol{\alpha})\}^2\right] = \varepsilon\left[e_1^2\right] \tag{5.8}$$

$$E_2 = \varepsilon\left[\{y_2(\boldsymbol{\chi}) - h(\boldsymbol{\chi})\}^2\right] = \varepsilon\left[e_2^2\right] \tag{5.9}$$

$$E_3 = \varepsilon\left[\{y_2(\boldsymbol{\kappa}) - h(\boldsymbol{\kappa})\}^2\right] = \varepsilon\left[e_3^2\right] \tag{5.10}$$

Here $\varepsilon[\bullet]$ indicates the expected value and corresponds to integration over the input data, and is defined as follows (Perrone and Cooper 1993):

$$\varepsilon\left[e_1^2\right] \equiv \int e_1^2(\alpha)\, p(\alpha)\, d\alpha \tag{5.11}$$

$$\varepsilon\left[e_2^2\right] \equiv \int e_2^2(\chi)\, p(\chi)\, d\chi \tag{5.12}$$

$$\varepsilon\left[e_3^2\right] \equiv \int e_3^2(\kappa)\, p(\kappa)\, d\kappa \tag{5.13}$$

Here $p[\bullet]$ is the probability density function; and $d[\bullet]$ is a differential operator. The average MSE of the three networks acting individually may be written as follows (Perrone and Cooper 1993):

$$\begin{aligned} E_{AV} &= \frac{E_1(\alpha) + E_2(\chi) + E_3(\kappa)}{3} \\ &= \frac{1}{3}\left(\varepsilon\left(e_1^2\right) + \varepsilon\left(e_2^2\right) + \varepsilon\left(e_3^2\right)\right) \end{aligned} \tag{5.14}$$

5.4.1 Equal Weights

In this section, the concept of a committee is explained. The output of the committee is the average of the outputs from the pseudo-modal energies, modal-property, and WT networks. The committee prediction may be written in the following form by giving equal weighting functions (Perrone and Cooper 1993):

$$y_{COM} = \frac{1}{3}\left(y_1(\alpha) + y_2(\chi) + y_3(\kappa)\right) \tag{5.15}$$

The MSE of the committee can therefore be written as follows:

$$\begin{aligned} E_{COM} &= \varepsilon\left[\left(\frac{1}{3}\left\{y_1(\alpha) + y_2(\chi) + y_3(\kappa)\right\} - \frac{1}{3}\left[h(\alpha) + h(\chi) + h(\kappa)\right]\right)^2\right] \\ &= \varepsilon\left[\left(\frac{1}{3}\left\{\left[y_1(\alpha) - h(\alpha)\right] + \left[y_2(\chi) - h(\chi)\right] + \left[y_3(\kappa) - h(\kappa)\right]\right\}\right)^2\right] \\ &= \varepsilon\left[\left(\frac{1}{3}\left\{e_1 + e_2 + e_3\right\}\right)^2\right] \\ &= \frac{1}{9}\left(\varepsilon\left[e_1^2\right] + 2\left(\varepsilon\left[e_1 e_2\right] + \varepsilon\left[e_2 e_3\right] + \varepsilon\left[e_1 e_3\right]\right) + \varepsilon\left[e_2^2\right] + \varepsilon\left[e_3^2\right]\right) \end{aligned} \tag{5.16}$$

If it is assumed that the errors (e_1, e_2, and e_3) are uncorrelated then

$$\varepsilon\left[e_1 e_2\right] = \varepsilon\left[e_1 e_3\right] = \varepsilon\left[e_2 e_3\right] = 0 \tag{5.17}$$

Substituting Eq. 5.17 in Eq. 5.16, the error of the committee can be related to the average error of the networks acting individually as follows (Perrone and Cooper 1993):

$$\begin{aligned} E_{COM} &= \frac{1}{9}\left(\varepsilon\left[e_1^2\right] + \varepsilon\left[e_2^2\right] + \varepsilon\left[e_3^2\right]\right) \\ &= \frac{1}{3} E_{AV} \end{aligned} \tag{5.18}$$

Equation 5.18 demonstrates that the MSE of the committee is one-third of the average MSE of the individual technique. From Eq. 5.18, it can be deduced that the MSE of the committee is always equal to or less than the average MSE of the three methods acting individually.

5.4.2 Variable Weights

The three networks might not essentially have the same predictive capability. This might be because modal properties are extracted from the FRFs (Ewins 1995), the WT of the impulse response that is chosen is not ideal, or the parameters chosen from the pseudo-modal energies are not ideal. To accommodate the strength of each technique, the network should be given suitable weighting functions. It will be explained later how these weighting functions will be evaluated when there is no prior knowledge of the strength of each approach.

The identity of fault may be defined as the combination of the three independent methods with estimated weighting functions as (a modification of Eq. 5.15):

$$y_{COM} = \gamma_1 y_1(\boldsymbol{\alpha}) + \gamma_2 y_2(\boldsymbol{\chi}) + \gamma_3 y_3(\boldsymbol{\kappa}) \tag{5.19}$$

Here γ_1, γ_2, and γ_3 are the weighting functions and $\gamma_1 + \gamma_2 + \gamma_3 = 1$. The MSE due to the weighted committee can be written as follows (Marwala 2000):

$$\begin{aligned} E_{COM} &= \varepsilon\left[\left(\gamma_1 y_1(\boldsymbol{\alpha}) + \gamma_2 y_2(\boldsymbol{\chi}) + \gamma_3 y_3(\boldsymbol{\kappa}) - \left[\gamma_1 h(\boldsymbol{\alpha}) + \gamma_2 h(\boldsymbol{\chi}) + \gamma_3 h(\boldsymbol{\kappa})\right]\right)^2\right] \\ &= \varepsilon\left[\left(\gamma_1\left[y_1(\boldsymbol{\alpha}) - h(\boldsymbol{\alpha})\right] + \gamma_2\left[y_2(\boldsymbol{\chi}) - h(\boldsymbol{\chi})\right] + \gamma_3\left[y_3(\boldsymbol{\kappa}) - h(\boldsymbol{\kappa})\right]\right)^2\right] \\ &= \varepsilon\left[\left(\gamma_1 e_1 + \gamma_2 e_2 + \gamma_3 e_3\right)^2\right] \end{aligned} \tag{5.20}$$

Equation 5.20 may be rewritten in Lagrangian form as follows (Perrone and Cooper 1993):

$$E_{COM} = \varepsilon\left[(\gamma_1 e_1 + \gamma_2 e_2 + \gamma_3 e_3)^2\right] + \lambda\,(1 - \gamma_1 - \gamma_2 - \gamma_3) \tag{5.21}$$

Here λ is the Lagrangian multiplier. The derivative of error in Eq. 5.21 with respect to γ_1, γ_2, γ_3 and λ may be calculated and equated to zero as (Perrone and Cooper 1993):

$$\frac{dE_{COM}}{d\gamma_1} = 2e_1\varepsilon\,[(\gamma_1 e_1 + \gamma_2 e_2 + \gamma_3 e_3)] - \lambda = 0 \tag{5.22}$$

$$\frac{dE_{COM}}{d\gamma_2} = 2e_2\varepsilon\,[(\gamma_1 e_1 + \gamma_2 e_2 + \gamma_3 e_3)] - \lambda = 0 \tag{5.23}$$

$$\frac{dE_{COM}}{d\gamma_3} = 2e_3\varepsilon\,[(\gamma_1 e_1 + \gamma_2 e_2 + \gamma_3 e_3)] - \lambda = 0 \tag{5.24}$$

$$\frac{dE_{COM}}{d\lambda} = 1 - \gamma_1 - \gamma_2 - \gamma_3 = 0 \tag{5.25}$$

In solving Eqs. 5.21–5.25, the minimum error is obtained when the weights are (Perrone and Cooper 1993):

$$\gamma_1 = \frac{1}{1 + \frac{\varepsilon[e_1^2]}{\varepsilon[e_2^2]} + \frac{\varepsilon[e_1^2]}{\varepsilon[e_3^2]}} \tag{5.26}$$

$$\gamma_2 = \frac{1}{1 + \frac{\varepsilon[e_2^2]}{\varepsilon[e_1^2]} + \frac{\varepsilon[e_2^2]}{\varepsilon[e_3^2]}} \tag{5.27}$$

$$\gamma_3 = \frac{1}{1 + \frac{\varepsilon[e_3^2]}{\varepsilon[e_1^2]} + \frac{\varepsilon[e_3^2]}{\varepsilon[e_2^2]}} \tag{5.28}$$

Equations 5.26–5.28 may be generalized for a committee with n-trained networks and may be written as follows (Perrone and Cooper 1993):

$$\gamma_i = \frac{1}{\sum_{j=1}^{n} \frac{\varepsilon[e_i^2]}{\varepsilon[e_j^2]}} \tag{5.29}$$

From Eq. 5.29, the following conditions may be derived (Marwala 2000):

$$\varepsilon\left[e_1^2\right] = \varepsilon\left[e_2^2\right] = \varepsilon\left[e_3^2\right] \Rightarrow \gamma_1 = \gamma_2 = \gamma_3 = \frac{1}{3} \tag{5.30}$$

$$\varepsilon\left[e_3^2\right] < \varepsilon\left[e_2^2\right] < \varepsilon\left[e_1^2\right] \Rightarrow \gamma_1 < \gamma_2 < \gamma_3; \gamma_3 > \frac{1}{3} \tag{5.31}$$

$$\varepsilon\left[e_1^2\right] < \varepsilon\left[e_2^2\right] < \varepsilon\left[e_3^2\right] \Rightarrow \gamma_3 < \gamma_2 < \gamma_3; \gamma_1 > \frac{1}{3} \tag{5.32}$$

These conditions show that if the predictive capacity of the pseudo-modal energy network, the modal-property network, and the WT network are equal, then each method should be given equal weights. If the pseudo modal energy network is better than the other two, it should be given a higher weight. If the modal-property network has lower expected errors than the other two networks, it should be given a higher weight. If the WT network has smaller errors than the other two methods, it should be given more weight. These conditions are trivial, but they have been derived to confirm the effectiveness of the presented technique.

Because it is not known which network is more accurate in a given instance, the weighting functions were determined from the data that was used for training the networks (prior knowledge) and this is called *stacking*, as described in the previous section.

Therefore, it can be concluded that if three independent (uncorrelated) methods are used simultaneously, the reliability of the combination is at least as good as when the methods are used individually. Suppose the probabilities of success for the pseudo-modal energy network, the modal-property network, and the WT network are $P(x_1)$, $P(x_2)$, and $P(x_3)$, respectively. The reliability of the three methods acting in parallel can be mathematically written as follows (Marwala 2001):

$$\begin{aligned} P\left(x_1 \cup x_2 \cup x_3\right) = & \, P\left(x_1\right) + P\left(x_2\right) + P\left(x_3\right) - \left[P\left(x_1 \cap x_2\right)\right. \\ & + P\left(x_2 \cap x_3\right) P\left(x_1 \cap x_3\right)\left.\right] \ldots \\ & + P\left(x_1 \cap x_2 \cap x_3\right) \end{aligned} \tag{5.33}$$

From Eq. 5.33, it can be concluded that the reliability of the committee is always higher than that of the individual methods.

5.5 Application to Cylindrical Shells

For this section, the committee procedure was applied to identify faults in a population of cylindrical shells. As described in Chap. 2, an impulse hammer test was performed on ten steel seam-welded cylindrical shells (1.75 ± 0.02 mm thickness, 101.86 ± 0.29 mm diameter, and height 101.50 ± 0.20 mm). These cylinders were resting on bubble wrap to simulate a free-free environment. The details of this experiment may be found in Marwala (1999).

Table 5.1 Confusion matrix from the classification of fault cases in the test data using the pseudo-modal-energy network

		Predicted							
		[000]	[100]	[010]	[001]	[110]	[101]	[011]	[111]
Actual	[000]	37	2	0	0	0	0	0	0
	[100]	0	3	0	0	0	0	0	0
	[010]	0	0	3	0	0	0	0	0
	[001]	0	0	0	3	0	0	0	0
	[110]	0	0	0	0	3	0	0	0
	[101]	0	0	0	0	0	3	0	0
	[011]	0	0	0	0	0	0	3	0
	[111]	0	0	0	0	5	1	1	32

Each cylinder was divided into three substructures, and holes of 12 mm diameter were drilled into each substructure. For one cylinder, the first type of fault was a zero-fault scenario, and its identity was [000]. The second type of fault was a one-fault scenario; if it was located in substructure 1, its identity was [100]. The third type of fault was a two-fault scenario, and if the faults were located in substructures 1 and 2, the identity of this case was [110]. The final type of fault was a three-fault scenario, and the identity of this case was [111].

For each fault case, measurements were taken by measuring the acceleration at a fixed position and roving the impulse position about. One cylinder gives four fault scenarios and 12 sets of measurements. The structure was vibrated at 19 different locations, nine on the upper ring of the cylinder and ten on the lower ring of the cylinder. Each measurement was taken three times to quantify the repeatability of the measurements. The total number of data points collected was 120.

From the measured data, pseudo-modal energies, modal properties, and wavelet data were identified and used to train three neural networks. The training process was done in the same way as in Chap. 3. The WT network was trained using wavelet data. This network had 18 input parameters, 9 hidden units, and 3 output units. The committee was applied using the weighting obtained from the validation data.

When the networks were evaluated using the data not used for training, the results in Tables 5.1, 5.2, 5.3 and 5.4 were obtained. These results indicate that the committee approach gave the best results followed by the pseudo-modal energy network, and then the modal-property network. The wavelet-network performed the worst.

The neural networks were trained using the hybrid Monte Carlo method as described in Chap. 4. The number of initial states discarded in the hope of reaching a stationary distribution was set to 100; the number of steps in each hybrid Monte Carlo trajectory was set to 100; the fixed step size was set to 0.001; and the number of samples retained to form a distribution was set to 500.

Table 5.2 Confusion matrix from the classification of fault cases in the test data using the modal-energy-network

		Predicted							
		[000]	[100]	[010]	[001]	[110]	[101]	[011]	[111]
Actual	[000]	38	0	0	1	0	0	0	0
	[100]	0	3	0	0	0	0	0	0
	[010]	0	0	3	0	0	0	0	0
	[001]	0	0	0	3	0	0	0	0
	[110]	0	0	0	0	3	0	0	0
	[101]	0	0	0	0	0	3	0	0
	[011]	0	0	0	0	0	0	3	0
	[111]	0	0	0	0	5	2	6	26

Table 5.3 Confusion matrix from the classification of fault cases in the test data using the wavelet-network

		Predicted							
		[000]	[100]	[010]	[001]	[110]	[101]	[011]	[111]
Actual	[000]	35	0	0	1	0	0	0	0
	[100]	0	1	0	0	0	0	0	0
	[010]	2	0	3	0	0	0	0	0
	[001]	0	2	0	2	0	0	0	0
	[110]	0	0	0	0	3	0	0	1
	[101]	0	0	0	1	0	3	0	0
	[011]	0	0	0	0	0	0	3	0
	[111]	1	0	0	0	5	2	6	25

Table 5.4 Confusion matrix from the classification of fault cases in the test data using the committee-network

		Predicted							
		[000]	[100]	[010]	[001]	[110]	[101]	[011]	[111]
Actual	[000]	38	1	0	0	0	0	0	0
	[100]	0	3	0	0	0	0	0	0
	[010]	0	0	3	0	0	0	0	0
	[001]	0	0	0	3	0	0	0	0
	[110]	0	0	0	0	3	0	0	0
	[101]	0	0	0	0	0	3	0	0
	[011]	0	0	0	0	0	0	3	0
	[111]	0	0	1	0	3	0	1	34

5.6 Conclusions

In this study, the committee of neural networks method was presented and applied to the structural diagnostics of a population of cylindrical shells. This method used pseudo-modal energies, modal properties, and used wavelet transform data to simultaneously identify faults in structures. It was observed that the committee approach gave the best results followed by the pseudo-modal energy network and then the modal-property network, while the wavelet-network performed the worst.

References

Abdel-Aal RE (2005a) Improving electric load forecasts using network committees. Electric Power Syst Res 74:83–94

Abdel-Aal RE (2005b) Improved classification of medical data using abductive network committees trained on different feature subsets. Comput Methods Program Biomed 80:141–153

Anthony M (2007) On the generalization error of fixed combinations of classifiers. J Comput Syst Sci 73:725–734

Atalla MJ, Inman DJ (1998) On model updating using neural networks. Mech Syst Signal Process 12:135–161

Bacauskiene M, Verikas A (2004) Selecting salient features for classification based on neural network committees. Pattern Recognit Lett 25:1879–1891

Baras D, Fine S, Fournier L, Geiger D, Ziv A (2011) Automatic boosting of cross-product coverage using bayesian networks. Int J Software Tools Technol Transfer 13:247–261

Basu M (2004) An interactive fuzzy satisfying method based on evolutionary programming technique for multiobjective short-term hydrothermal scheduling. Electric Power Syst Res 69:277–285

Bichindaritz I, Annest A (2010) Case based reasoning with bayesian model averaging: an improved method for survival analysis on microarray data. Lect Notes Comput Sci 6176:346–359

Bishop CM (1995) Neural networks for pattern recognition. Oxford University Press, Oxford

Bobb JF, Dominici F, Peng RD (2011) A bayesian model averaging approach for estimating the relative risk of mortality associated with heat waves in 105 U.S. cities. Biometrics 67:1605–1616

Boone EL, Ye K, Smith EP (2011) Assessing environmental stressors via Bayesian model averaging in the presence of missing data. Environmetrics 22:13–22

Breiman L (1996) Bagging predictors. Mach Learn 24:123–140

Cao YJ, Jiang L, Wu QH (2000) An evolutionary programming approach to mixed-variable optimization problems. Appl Math Model 24:931–942

Chen CH, Lin ZS (2006) A committee machine with empirical formulas for permeability prediction. Comput Geosci 32:485–496

Clarke B (2003) Comparing Bayes model averaging and stacking when model approximation error cannot be ignored. J Mach Learn Res 4:683–712

Das A, Reddy NP, Narayanan J (2001) Hybrid fuzzy logic committee neural networks for recognition of swallow acceleration signals. Comput Methods Programs Biomed 64:87–99

Daubechie I (1991) The wavelet transform, time-frequency localization and signal processing. IEEE Trans Info Theory 36:961–1005

Doebling SW, Farrar CR, Prime MB, Shevitz DW (1996) Damage identification and health monitoring of structural and mechanical systems from changes in their vibration characteristics: a literature review. Los Alamos National Laboratory report LA-13070-MS, Los Alamos National Laboratory, Los Alamos

Domingos P (2000) Bayesian averaging of classifiers and the overfitting problem. In: Proceedings of the 17th international conference on machine learning, Stanford, California, pp 223–230

Drygajlo A, Li W, Qiu H (2011) Adult face recognition in score-age-quality classification space. Lect Notes Comput Sci 6583:205–216

Du J, Zhai C, Wan Y (2007) Radial basis probabilistic neural networks committee for palmprint recognition. Lect Notes Comput Sci 4492:819–824

Ewins DJ (1995) Modal testing: theory and practice. Research Studies Press, Letchworth
Feldkircher M (2011) Forecast combination and Bayesian model averaging: a prior sensitivity analysis. J Forecast doi:10.1002/for.1228
Fernandes AM, Utkin AB, Lavrov AV, Vilar RM (2004) Development of neural network committee machines for automatic forest fire detection using Lidar. Pattern Recognit 37:2039–2047
Friswell MI, Mottershead JE (1995) Finite element model updating in structural dynamics. Kluwer Academic Publishers, Dordrecht
Hanczar B, Nadif M (2011) Using the bagging approach for biclustering of gene expression data. Neurocomputing 74:1595–1605
Haussler D, Kearns M, Schapire RE (1994) Bounds on the sample complexity of Bayesian learning using information theory and the VC dimension. Mach Learn 14:83–113
Hernandez-Lobato D, Martinez-Munoz G, Suarez A (2011) Empirical analysis and evaluation of approximate techniques for pruning regression bagging ensembles. Neurocomputing 74: 2250–2264
Hoeting JA, Madigan D, Raftery AE, Volinsky CT (1999) Bayesian model averaging: a tutorial. Stat Sci 14:382–401
Homayouni H, Hashemi S, Hamzeh A (2010) Instance-based ensemble learning algorithm with stacking framework. In: Proceedings of the 2nd international conference on software technology and engineering, Puerto Rico, USA, pp 164–169
Hu G, Mao Z, He D, Yang F (2011) Hybrid modeling for the prediction of leaching rate in leaching process based on negative correlation learning bagging ensemble algorithm. Comput Chem Eng 35(12):2611–2617
Huang X, Zhang L, Gong W (2011) Information fusion of aerial images and LIDAR data in urban areas: vector-stacking, re-classification and post-processing approaches. Int J Remote Sens 32:69–84
Imregun M, Visser WJ, Ewins DJ (1995) Finite element model updating using frequency response function data I: theory and initial investigation. Mech Syst Signal Process 9:187–202
Jafari SA, Mashohor S, Jalali Varnamkhasti M (2011) Committee neural networks with fuzzy genetic algorithm. J Petrol Sci Eng 76:217–223
Jasra A, Holmes CC (2011) Stochastic boosting algorithms. Stat Comput 21:335–347
Jia J, Xiao X, Liu B, Jiao L (2011) Bagging-based spectral clustering ensemble selection. Pattern Recognit Lett 32:1456–1467
Jordan MI, Bishop CM (1996) Neural networks. MIT technology report artificial intelligence. Memo no. 1562, Massachusetts Institute of Technology, Cambridge
Kadkhodaie-Ilkhchi A, Reza Rezaee M, Rahimpour-Bonab H (2009) A committee neural network for prediction of normalized oil content from well log data: an example from South Pars Gas Field, Persian Gulf. J Petrol Sci Eng 65:23–32
Kajdanowicz T, Kazienko P (2011) Structured output element ordering in boosting-based classification. Lect Notes Comput Sci 6679:221–228
Karimpouli S, Fathianpour N, Roohi J (2010) A new approach to improve neural networks' algorithm in permeability prediction of petroleum reservoirs using Supervised Committee Machine Neural Network (SCMNN). J Petrol Sci Eng 73:227–232
Khoshgoftaar TM, van Hulse J, Napolitano A (2011) Comparing boosting and bagging techniques with noisy and imbalanced data. IEEE Trans Syst Man Cybern A Syst Hum 41:552–568
Kyung YJ, Lee HS (1999) Bootstrap and aggregating VQ classifier for speaker recognition. Electron Lett 35:973–974
Larios N, Lin J, Zhang M, Lytle D, Moldenke A, Shapiro L, Dietterich T (2011) Stacked spatial-pyramid kernel: an object-class recognition method to combine scores from random trees. In: Proceedings of the IEEE workshop on application of computer vision, Kona, Hawaii, pp 329–335
Leitenstorfer F, Tutz G (2011) Estimation of single-index models based on boosting techniques. Stat Model 11:203–217
Levin RI, Lieven NAJ (1998) Dynamic finite element updating using neural networks. J Sound Vib 210:593–608

Li W, Drygajlo A, Qiu H (2010) Aging face verification in score-age space using single reference image template. In: Proceedings of the IEEE 4th international conference on biometrics: theory, application and Systems, Washington, DC, pp 1–7

Liew KM, Wang Q (1998) Application of wavelet theory for crack identification in structures. J Eng Mech 124:152–157

Louzada F, Anacleto-Junior O, Candolo C, Mazucheli J (2011) Poly-bagging predictors for classification modelling for credit scoring. Expert Syst Appl 38:12717–12720

Maia NMM, Silva JMM (1997) Theoretical and experimental modal analysis. Research Studies Press, Letchworth

Marwala T (1999) Probabilistic damage identification using neural networks and modal properties. University of Cambridge Technical Report CUED/C-MECH/TR-76, University of Cambridge, Cambridge

Marwala T (2000) On damage identification using a committee of neural networks. J Eng Mech 126:43–50

Marwala T (2001) Probabilistic fault identification using a committee of neural networks and vibration data. J Aircraft 38:138–146

Marwala T (2009) Computational intelligence for missing data imputation, estimation and management: knowledge optimization techniques. IGI Global Publications, New York

Marwala T, Heyns PS (1998) Multiple-criterion method for determining structural damage. Am Inst Aeronaut Astronaut J 36:1494–1501

Marwala T, Hunt HEM (1999) Fault identification using finite element models and neural networks. Mech Syst Signal Process 13:475–490

Marwala T, de Wilde P, Correia L, Mariano P, Ribeiro R, Abramov V, Szirbik N, Goossenaerts J (2001) Scalability and optimisation of a committee of agents using genetic algorithm. In: Proceedings of the international symposium on soft computing and intelligent systems for industry, Paisley, Scotland, pp 1–6

Newland DE (1993) An introduction to random vibration, spectral and wavelet analysis. Longman/John Wiley, New York/Harlow

Osawa T, Mitsuhashi H, Uematsu Y, Ushimaru A (2011) Bagging GLM: improved generalized linear model for the analysis of zero-inflated data. Ecol Info 6(5):270–275

Park I, Grandhi RV (2010) Quantification of multiple types of uncertainty in computer simulation using bayesian model averaging. In: Proceedings of the 51st AIAA/ASME/ASCE/AHS/ASC structures, structural dynamics and materials conference, Orlando, Florida, pp 1–6

Park I, Grandhi RV (2011) Quantifying multiple types of uncertainty in physics-based simulation using Bayesian model averaging. Am Inst Aeronaut Astronaut J 49:1038–1045

Paya BA, Esat II, Badi MNM (1997) Artificial neural network based fault diagnostics of rotating machinery using wavelet transforms as a pre-processor. Mech Syst Signal Process 11:751–765

Perrone MP, Cooper LN (1993) When networks disagree: ensemble methods for hybrid neural networks. In: Mammone RJ (ed) Artificial neural networks for speech and vision. Chapman and Hall, London

Pino-Mejias R, Jimenez-Gamero MD, Cubiles-de-la-Vega MD, Pascual-Acosta A (2008) Reduced bootstrap aggregating of learning algorithms. Pattern Recognit Lett 29:265–271

Potempski S, Galmarini S, Riccio A, Giunta G (2010) Bayesian model averaging for emergency response atmospheric dispersion multimodel ensembles: is it really better? How many data are needed? Are the weights portable? J Geophys Res. doi:10.1029/2010JD014210

Rajan CCA, Mohan MR (2007) An evolutionary programming based simulated annealing method for solving the unit commitment problem. Int J Electric Power Energy Syst 29:540–550

Reddy NP, Buch OA (2003) Speaker verification using committee neural networks. Comput Methods Programs Biomed 72:109–115

Rokach L (2010) Ensemble-based classifiers. Artif Intell Rev 33:1–39

Sheikh-Ahmad J, Twomey J, Kalla D, Lodhia P (2007) Multiple regression and committee neural network force prediction models in milling FRP. Mach Sci Technol 11:391–412

Shi L, Xu G (2001) Self-adaptive evolutionary programming and its application to multi-objective optimal operation of power systems. Electric Power Syst Res 57:181–187
Shiraishi Y, Fukumizu K (2011) Statistical approaches to combining binary classifiers for multi-class classification. Neurocomputing 74:680–688
Sill J, Takacs G, Mackey L, Lin D (2009) Feature-weighted linear stacking. arXiv:0911.0460
Smyth P, Wolpert DH (1999) Linearly combining density estimators via stacking. Mach Learn J 36:59–83
Tang B, Chen Q, Wang X, Wang X (2010) Reranking for stacking ensemble learning. Lect Notes Comput Sci 6443:575–584
Tsai MY, Hsiao CK, Chen WJ (2011) Extended bayesian model averaging in generalized linear mixed models applied to schizophrenia family data. Ann Hum Genet 75:62–77
van Hinsbergen CPIJ, van Lint JWC, van Zuylen HJ (2009) Bayesian committee of neural networks to predict travel times with confidence intervals. Trans Res C Emerg Technol 17:498–509
Wang G, Hao J, Ma J, Jiang H (2011) A comparative assessment of ensemble learning for credit scoring. Expert Syst Appl 38:223–230
Wolpert DH (1992) Stacked generalization. Neural Netw 5:241–259
Wolpert DH, Macready WG (1999) An efficient method to estimate bagging's generalization error. Mach Learn J 35:41–55
Yu Q (2011) Weighted bagging: a modification of adaboost from the perspective of importance sampling. J Appl Stat 38:451–463
Zhao ZQ, Huang DS, Sun BY (2004) Human face recognition based on multi-features using Neural Networks Committee. Pattern Recognit Lett 25:1351–1358

Chapter 6
Gaussian Mixture Models and Hidden Markov Models for Condition Monitoring

6.1 Introduction

Rotating machines are widely used in industry for system operation and process automation. Research shows that the failures of these machines are often linked with bearing failures (Lou et al. 2004). Bearing faults induce high bearing vibrations which generate noise that may even cause the entire rotating machine, such as the electric motor, to function incorrectly. Thus, it is important to include bearing vibration fault detection and diagnosis in industrial motor rotational fault diagnosis systems (Lou et al. 2004). As a result, there is a high demand for cost effective automatic monitoring of bearing vibrations in industrial motor systems.

A variety of fault bearing vibration feature detection techniques exist. These can be classified into three domains, namely: frequency domain analysis, time-frequency domain analysis, and time domain analysis (Ericsson et al. 2004). The frequency domain methods often involve frequency analysis of the vibration signals and look at the periodicity of high frequency transients. This procedure is complicated by the fact that this periodicity may be suppressed (Ericsson et al. 2004). The most commonly used frequency analysis technique for detection and diagnosis of bearing fault is the *envelope analysis*. More details on this technique are found in McFadden and Smith (1984). The main disadvantage of the frequency domain analysis is that it tends to average out transient vibrations and therefore becomes more sensitive to background noise. To overcome this problem, the time-frequency domain analysis is used, which shows how the frequency contents of the signal changes with time. Examples of such analyses are: Short Time Fourier Transform (STFT), the Wigner-Ville Distribution (WVD) and, most importantly, the Wavelet Transform (WT). These techniques are studied in detail in the work of Li et al. (2000).

The last category of the feature detection is the time domain analysis. There are a number of time domain methods that give reasonable results. These methods include the time-series averaging method, the signal enveloping method, the Kurtosis method, and others (Li et al. 2000). Research shows that, unlike the frequency

T. Marwala, *Condition Monitoring Using Computational Intelligence Methods: Applications in Mechanical and Electrical Systems*, DOI 10.1007/978-1-4471-2380-4_6,

domain analysis, this technique is less sensitive to suppressions of the impact of periodicity (Ericsson et al. 2004; Li et al. 2000). This chapter introduces a new time domain analysis method, known as *fractal dimension analysis*, which was originally used in image processing and has been recently used in speech recognition (Maragos and Sun 1993; Maragos and Potamianos 1999; Wang et al. 2000). This method is expected to give enormous improvement to the performance of the bearing fault detection and diagnosis because it extracts the non-linear vibration features of each bearing fault. The fractal dimension analysis is based on the Multi-scale Fractal Dimensions (MFD) of short-time bearing vibration segments, derived from non-linear theory (Wang et al. 2000).

Once the bearing vibration features are extracted using one of the three domains mentioned above, then these features can be used for automatic motor bearing fault detection and diagnosis by applying them to a non-linear pattern classifier. The most popular classifier used in bearing fault detection is a Neural Network (NN). Nevertheless, other non-linear classifiers like Gaussian Mixture Model (GMM) and Hidden Markov Model (HMM) have been shown to outperform NN in a number of classification problems, in general, and in speech related problems in particular. Only recently, have researchers such as Purushothama et al. (2005) applied speech pattern classifiers, such as HMM, to the fault detection of mechanical systems because of their success in speech recognition.

This chapter presents a comparative study of HMM and GMM, and introduces time-domain analysis based techniques using fractals to extract the features. Furthermore, the ability of MFD to detect bearing faults was evaluated using both HMM and GMM non-linear pattern classifiers.

The rest of the chapter is arranged as follows: the next section presents the different bearing faults studied in this chapter, followed by the mathematical background to fractal dimensions, HMM, and GMM. Thereafter, the time domain bearing detection and diagnosis framework is presented.

6.2 Background

This section presents, in detail, the different bearing faults studied in this chapter, followed by the mathematical background to fractal dimensions, HMM, and GMM.

6.2.1 The Gaussian Mixture Model (GMM)

A GMM is a weighted sum of M component Gaussian densities, $p(\boldsymbol{x}|\lambda)$ as given by the equation (Reynolds 1992; Dempster et al. 1977):

$$p(\mathbf{x}|\lambda) = \sum_{i=1}^{M} w_i p_i(\mathbf{x}) \tag{6.1}$$

with

$$p_i(\mathbf{x}_t) = \frac{1}{(2\pi)^{D/2}\sqrt{\Sigma_i}} \exp\left\{-\frac{1}{2}(\mathbf{x}_k - \boldsymbol{\mu}_i)^t(\Sigma_i)^{-1}(\mathbf{x}_k - \boldsymbol{\mu}_i)\right\} \quad (6.2)$$

Here, $\boldsymbol{x}$ is a D-dimensional, continuous-valued data vector representing measurements from features, w_i, $i = 1, \ldots, M$, are the mixture weights, with mean vector μ_i and covariance matrix Σ_i. The mixture weights, w_i, satisfy the constraint $\sum_{i=1}^{M} w_i = 1$.

The entire GMM is parameterized by the mean vectors, covariance matrices, and mixture weights from all component densities and these parameters are together represented by the notation (Reynolds and Rose 1995; Dempster et al. 1977):

$$\lambda = \{\mathbf{w}, \boldsymbol{\mu}, \Sigma\} \quad (6.3)$$

Here, λ is the model, $\boldsymbol{w}$, $\boldsymbol{\mu}$, Σ are, respectively, the weights, means, and covariance of the features. The covariance matrices can be full rank or constrained to be diagonal but, in this chapter assumes that it is diagonal. The choice of model architecture, which are the number of components, diagonal covariance matrices and parameter is usually determined by the amount of data available for approximating the GMM parameters and how the GMM is applied in a specific fault identification problem. GMM has the advantage of being able to represent a large class of sample distributions and to form smooth estimates to indiscriminately shaped probability densities.

Given a collection of training vectors, the parameters of this model are estimated by a number of algorithms such as the *Expectation-Maximization (EM)* algorithm and K-means algorithm (Dempster et al. 1977; Reynolds et al. 2000). The EM algorithm was used in this study because it has reasonably fast computational time when compared to other algorithms. The EM algorithm finds the optimum model parameters by iteratively refining GMM parameters to increase the likelihood of the estimated model for the given bearing fault feature vector. More details on the EM algorithm for training a GMM are in the work of Wang and Kootsookos (1998).

Bordes et al. (2007) applied the EM algorithm to image reconstruction. They found that the results were within 10% of the experimental data. Dempster et al. (1977) applied the EM algorithm to missing data, while Ingrassia and Rocci (2007) generalized the EM algorithm to semi-parametric mixture models that, when tested on real data, showed that their method was easy to implement and computationally efficient. Kauermann et al. (2007) used the EM algorithm to recognize polymorphism in pharmacokinetic/pharmacodynamic (PK/PD) phenotypes, while Wang and Hu (2007) improved the EM algorithm's computational load and successfully applied this to brain tissue segmentation. Another successful implementation of the EM algorithm includes binary text classification (Park et al. 2007). Other improvements of the EM algorithm include accelerating the computational speed by Patel et al. (2007). Further information on the implementation of the EM algorithm can be found in Wang et al. (2007), as well as McLachlan and Krishnan (1997).

The aim of maximum likelihood estimation is to identify the model parameters which maximize the likelihood of the GMM, given the training data. For a series of T training vectors $X = \{x_1, \ldots, x_T\}$, the GMM likelihood, assuming independence between the vectors, can be expressed as (Reynolds 1992):

$$p(\mathbf{X}, \lambda) = \prod_{t=1}^{T} p(\mathbf{x}_t, \lambda) \tag{6.4}$$

For the EM algorithms, the re-estimations are calculated until convergence; and the mixture of weights, means, and variances can, respectively, be written as (Reynolds 1992):

$$\overline{w}_i = \frac{1}{T} \sum_{t=1}^{T} P(i\,|\mathbf{x}_t, \lambda) \tag{6.5}$$

$$\overline{\mu}_i = \frac{\sum_{t=1}^{T} P(i\,|\mathbf{x}_t, \lambda)\mathbf{x}_t}{\sum_{t=1}^{T} P(i\,|\mathbf{x}_t, \lambda)} \tag{6.6}$$

$$\overline{\sigma}_i^2 = \frac{\sum_{t=1}^{T} P(i\,|\mathbf{x}_t, \lambda)\mathbf{x}_i^2}{\sum_{t=1}^{T} P(i\,|\mathbf{x}_t, \lambda)} - \overline{\mu}_i^2 \tag{6.7}$$

The posterior probability can thus be written as (Reynolds 1992):

$$P(i\,|\mathbf{x}_t, \lambda) = \frac{w_i\, p(\mathbf{x}_t\,|\boldsymbol{\mu}_i, \Sigma_i)}{\sum_{k=1}^{M} w_k\, p(\mathbf{x}_t\,|\boldsymbol{\mu}_i, \Sigma_i)} \tag{6.8}$$

The bearing fault detection or diagnosis using this classifier is then achieved by computing the likelihood of the unknown vibration segment of the different fault models. This likelihood is given by (Dempster et al. 1977):

$$\hat{s} = \arg \max_{1 \le f \le F} \sum_{k=1}^{K} \log \mathrm{p}(\mathbf{x}_k|\lambda_f) \tag{6.9}$$

Here F represents the number of faults to be diagonalized, $X = \{x_1, x_2, \ldots, x_K\}$ is the unknown D-dimension bearing fault-vibration segment.

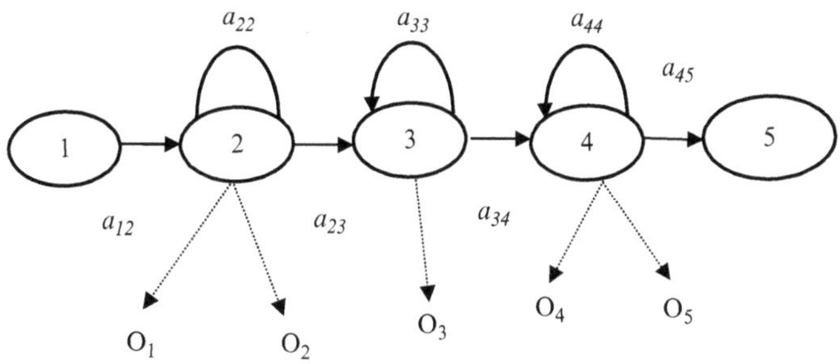

Fig. 6.1 Markov chain with five states with selected state transitions. Here O is the observation and *a* is the transition probability

6.2.2 The Hidden Markov Model (HMM)

The HMM is a statistical Markov model in which the system being modeled is assumed to be a Markov process with states that are hidden and therefore cannot be observed. In a conventional Markov model, the state is observable, and consequently, the transition probabilities are the only parameters to be estimated while naturally, the output is visible.

Essentially, HMM is a stochastic signal model. HMMs are referred to as Markov sources or probabilistic functions of Markov chains (Rabiner 1989). This model has been applied mostly to speech recognition systems and only recently it has been applied to bearing fault detection. In HMM, the observation is a probabilistic function of the state and this means the resulting model is a doubly emended stochastic process with an underlining stochastic process that is not observable (Rabiner 1989). Nevertheless, this process can only be observed through another stochastic process that produces the sequence. There are a number of possible Markov models, but the left-to-right model is typically applied in speech recognition. The structure of this model is shown in Fig. 6.1 with five states (Rabiner 1989).

Marwala et al. (2006) used bearing vibration signals features which were extracted using a time-domain fractal-based feature extraction technique as well as the HMM and GMM for bearing fault detection. The fractal technique was the Multi-Scale Fractal Dimension and was estimated using the Box-Counting Dimension. The extracted features were then applied to classify faults using the GMM and HMM. The results showed that the HMM outperformed the GMM and that the HMM was computationally more expensive than the GMM.

Boutros and Liang (2011) applied the discrete HMM for the detection and diagnosis of bearing and cutting tool faults. Their method was tested and validated using two situations, tool fracture, and bearing faults. In the first situation, the model correctly detected the state of the tool and, in the second case; the model classified the severity of the fault seeded into two different engine bearings. The result obtained for fault severity classification was above 95%. In addition to the fault severity, a location index was developed to determine the fault location and gave an average success rate of 96%.

Wong and Lee (2010) successfully applied HMM for fault detection in the shell-and-tube heat exchanger. This method was viewed as a generalization of the mixture-of-Gaussians method and was demonstrated through a problem.

Lee et al. (2010) applied HMM for online degradation assessment and adaptive fault detection of multiple failure modes. Their method, together with statistical process control was used to detect the incidence of faults. This technique permitted the hidden Markov state to be updated with the identification of new states. The results for a turning process showed that the tool wear processes can be successfully detected, and the tool wear processes can be identified.

Calefati et al. (2006) successfully applied HMM for machine faults detection and forecasting in gearboxes. Elsewhere, Zhou and Wang (2005) applied HMM and a principal component analysis to the on-line fault detection and diagnosis in industrial processes, and applied these to case studies from the Tennessee Eastman process.

Menon et al. (2003) applied HMM for incipient fault detection and diagnosis in turbine engines and the effectiveness of the HMM method was compared to a neural network method and a hybrid of principal component analysis and a neural network approach. Their HMM method was found to be more effective than the other methods.

Smyth (1994) applied HMM to fault detection in dynamic systems. It was demonstrated that a pattern recognition system combined with a finite-state HMM was good at modeling temporal characteristics. The model was validated using a real-world fault diagnosis problem and was demonstrated to offer substantial practical advantages.

The complete parameter set needed to define the HMM can be written as (Rabiner 1989; Caelli et al. 2001; Koski 2001):

$$\lambda = \{A, B, \pi\} \tag{6.10}$$

where λ is the model, $A = \{a_{ij}\}$, $B = \{b_{ij}(k)\}$ and $\pi = \{\pi_i\}$are the transition probability distribution, the observation probability distribution, and initial state distribution, respectively. For example, if we assume that the distribution can be represented by the Gaussian mixture model shown in Eq. 6.2, the equation can be written as:

$$\lambda = \{A, \mathbf{w}, \boldsymbol{\mu}, \boldsymbol{\Sigma}, \pi\} \tag{6.11}$$

These parameters of a given state, S_i, are defined as (Rabiner 1989; Ching et al. 2003; Purushothama et al. 2005; Ching and Ng 2006):

$$a_{ij} = P(q_{t+1} = S_j | q_t = S_i), 1 \leq i, j \leq N \tag{6.12}$$

$$b_{ij}(k) = P(O_k | q_t = S_i), 1 \leq j \leq N, 1 \leq k \leq M \tag{6.13}$$

and

$$\pi_i = P(q_1 = S_i), 1 \leq i \leq N \tag{6.14}$$

Here, q_t is the state at time t and N denotes the number of states. Additionally, $O_k$is the kth observation and M is the number of distinct observation.

The HMM can be used to simulate the observed state as follows (Rabiner 1989):

1. Let $t = 1$.
2. Create $O_t = v_k \in V$ in accordance with the probability $b_i(k)$.
3. Create a transition of hidden state from $q_t = S_i$ to $q_{t+1} = S_j$ in accordance with the transition probability a_{ij}.
4. Let $t = t + 1$ and go to Step 2 if $t < T$ or else terminate the algorithm.

There are three fundamental issues to be solved for this model to be applied in practice. Firstly, we ought to identify the probability of the observation sequence $\mathbf{O} = O_1, O_2, ..., O_T$ of visible states generated by the model λ. Secondly, we need a decoding process which identifies a state sequence that maximizes the probability of an observation sequence and this can be realized through the so-called Viterbi algorithm (Rabiner 1989). Thirdly, we need a training process which adjusts model parameters to maximize the probability of the observed sequence.

The next step is to calculate the likelihood of the observed sequence as follows (Rabiner 1989; Ching et al. 2004):

$$P(\mathbf{O}) = \sum_{all\ possible} \pi_{q_1} b_{q_1}(O_1) \times \pi_{q_2} b_{q_2}(O_2) \times ... \times \pi_{q_n} b_{q_n}(O_n) \tag{6.15}$$

To speed up the computation of this, the backward and the forward methods can be applied (Baum 1972). To do this, we define the following (Baum 1972):

$$\alpha_T(i) = P(O_1 O_2 ... O_t, q_t = S_i) \tag{6.16}$$

The forward technique can be written as follows (Rabiner 1989; Tai et al. 2009):

1. Initialize as follows:

$$\alpha_1(i) = \pi_i b_i(O_1) for\ 1 \leq i \leq N$$

2. Apply the recursion step as follows:

$$\alpha_t(j) = b_j(O_t)\sum_{i=1}^{N}\alpha_{t-1}(i)a_{ij}\; for\, 2 \le t \le T\; and\, 1 \le j \le N$$

3. Terminate as follows:

$$P(\mathbf{O}) = \sum_{i=1}^{N}\alpha_T(i)$$

The backward technique can be written as follows by letting (Rabiner 1989; Tai et al. 2009):

$$\beta_t(i) = P(O_{t+1}O_{t+2}...O_T\,|q_t = S_i) \tag{6.17}$$

1. Initialize as follows:

$$\beta_T(i) = 1\, for\, 1 \le i \le N$$

2. Apply the recursion step as follows:

$$\beta_t(i) = \sum_{j=1}^{N} a_{ij}b_j(O_{t+1})\beta_{t+1}(j)\, for\; 1 \le t \le T-1\; and\; 1 \le j \le N$$

3. Terminate as follows:

$$P(\mathbf{O}) = \sum_{i=1}^{N}\beta_1(i)\pi_i b_i(O_1)$$

The Baum-Welch estimation procedures with the Maximum Likelihood technique can be used to approximate the model parameters (Rabiner 1989). To explain the use of this procedure it is important to state the following definition (Rabiner 1989; Ching et al. 2004):

$$\xi_t(i,j) = P\left(q_t = S_i, q_{t+1} = S_j\,|O, A, B, \pi\right) \tag{6.18}$$

This is the probability of being in state S_i at time t and having a transition to state S_i at time $t+1$ given the observed sequence and the model. This can be expanded as follows (Rabiner 1989; Ching et al. 2004):

$$\xi_t(i,j) = \frac{\alpha_t(i)\alpha_{ij}\beta_{t+1}(j)b_j(O_{t+1})}{\sum_i \sum_j \alpha_t(i)\alpha_{ij}\beta_{t+1}(j)b_j(O_{t+1})} \tag{6.19}$$

We can also define that (Rabiner 1989; Ching et al. 2004):

$$\gamma_t(i) = P(q_t = S_i \,|O, A, B, \pi) \tag{6.20}$$

This indicates the probability of being in state S_i at time t given the observed sequence and the model. Therefore, we now have (Rabiner 1989; Ching et al. 2004):

$$\gamma_t = \sum_j \xi_t(i,j) \tag{6.21}$$

This procedure can be written as follows (Baum 1972; Rabiner 1989):

1. Select a set of initial parameters $\lambda = \{A, B, \pi\}$ randomly
2. Estimate the parameters using the following equations (Tai et al. 2009)

$$\bar{\pi}_i = \gamma_1(i) \; for \; 1 \leq i \leq N$$

$$\bar{a}_{ij} = \frac{\sum_{t=1}^{T-1} \xi_t(i,j)}{\sum_{t=1}^{T-1} \gamma_t(i)} \; for \; 1 \leq i \leq N, 1 \leq j \leq N$$

$$\bar{b}_j(k) = \frac{\sum_{t=1}^{T} \gamma_t(j) I_{O_t=k}}{\sum_{t=1}^{T} \gamma_t(j)} \; for \; 1 \leq j \leq N, 1 \leq k \leq M$$

Here $I_{O_t=k} = \begin{cases} 1 & if \; O_t = k \\ 0 & otherwise \end{cases}$

3. Set $\bar{A} = \{\bar{a}_{ij}\}_{ij}, \bar{B} = \{\bar{b}_j(k)\}_{jk} \; and \; \bar{\pi} = \{\bar{\pi}_i\}$
4. Set $\bar{\lambda} = \{\bar{A}, \bar{B}, \bar{\pi}\}$
5. If $\lambda = \bar{\lambda}$, end otherwise let $\lambda = \bar{\lambda}$ and go to Step 2

A more detailed explanation of HMM training using the Baum-Welch re-estimation along with other features of HMM is presented by Rabiner (1989).

The estimation of the hidden state can be conducted using the Viterbi algorithm (Viterbi 1967) to calculate the probability of the hidden states given the HMM parameters and an observed sequence. To do this we can define the following (Rabiner 1989; Tai et al. 2009) which is the maximum probability within a single path:

$$\delta_t(i) = \max_{q_1, q_2, \ldots, q_{t-1}} P\left(q_1, q_2, \ldots, q_t, O_1 O_2 \ldots O_t; q_t = S_i\right) \tag{6.22}$$

and define (Tai et al. 2009):

$$\delta_t(j) = b_j(O_t) \times \max_i\{\delta_{t-1}(i)a_{ij}\} \tag{6.23}$$

We can therefore solve this problem using dynamic programming as follows (Rabiner 1989; Tai et al. 2009):

1. Initialize $\delta_1(i) = \pi_i b_i(O_1)$ *and* $\theta_1(i) = 0$ *for* $1 \le i \le N$
2. Solve the following recursion step

$$\delta_t(j) = \max_{1\le i\le N} \delta_{t-1}(i)a_{ij}b_j(O_t)\,for\; 2 \le t \le T\; and\; 1 \le j \le N$$

and

$$\theta_t(j) = \arg\max_{1\le i\le N}\{\delta_{t-1}(i)a_{ij}\}\,for\; 2 \le t \le T\; and\; 1 \le j \le N$$

3. Terminate

$$P^* = \max_{1\le i\le N} \delta_T(i)\; and\; q_T^* = \arg\max_{1\le i\le N} \delta_T(i)$$

Here P^* is the most likely likelihood and q^* is the most likely state at time T.
4. Backtrack:

$$q_T^* = \theta_{t+1}(q_{t+1}^*)\,for\; t = T-1, T-2, \dots, 2, 1$$

6.2.3 Fractals

For this chapter, fractals were used to analyse the bearing data. A *fractal* is defined as a rough geometric shape that can be divided into various segments, each of which is roughly a reduced-size copy of the whole. This characteristic is known as self-similarity (Mandelbrot 1982). The theory of fractals was described in detail, in Chap. 2. The basis of the idea of fractals extends back to the seventeenth century. There are numerous classes of fractals, characterized as displaying exact self-similarity, quasi self-similarity or statistical self-similarity (Briggs 1992). Even though fractals are a mathematical concept, they are seen in nature, and this has led to their use in the arts they are useful in biomedical sciences, engineering sciences, and speech recognition. A fractal usually has the following characteristics (Falconer 2003):

- It contains a fine structure at randomly small scales.
- It is too irregular to be described using Euclidean geometry.
- It is approximately self-similar.

- It has a Hausdorff dimension (this explained in Chap. 2) which is more than its topological dimension (Pickover 2009).
- It has a basic and recursive description.

Fractals are frequently viewed to be infinitely complex because they look similar at all levels of magnification (Batty 1985; Russ 1994). Examples of natural objects that can be approximated by fractals include clouds, mountain ranges, lightning bolts, and coastlines (Sornette 2004).

Zhang et al. (2010) successfully applied a combined wavelet and fractal method for the fault detection of the opening fault of power electronic circuits based on the singularity of the fault signal from the power electronic equipment. Voltage wave signals were analyzed by applying the wavelet transform and correlative dimensions of the wavelet transform were estimated using fractals.

Yang et al. (2011) applied a fractal correlation dimension for the fault detection in the supply air temperature sensors of air handling unit systems and the results obtained demonstrated that it was more efficient in detecting a relatively small bias fault under noise conditions.

Ikizoglu et al. (2010) applied a Hurst parameter and fractal dimension for fault the detection of the bearings in electric motors. The vibration signals were obtained, analyzed in the frequency domain.

Ma (2009) successfully applied fractal analysis for fault detection in the welding process while Shanlin et al. (2007) successfully applied wavelet fractal network for fault detection in a power system generator.

Other successful applications of the wavelet transform in fault detection include its application for distributed power system short circuit problems (Song et al. 2007), robotics (Yan et al. 2007), short-circuit faults in low-voltage systems (Kang et al. 2006), and Direct Current system grounding (Li et al. 2005).

6.3 Motor Bearing Faults

Vibration measurement is important in advanced conditioning monitoring of mechanical systems. Most bearing vibrations are periodical movements. In general, rolling bearing contains two concentric rings, which are called the inner and outer raceway and these were shown in Chap. 2 (Li et al. 2000). Furthermore, the bearing contains a set of rolling elements that run in the tracts of these raceways. There is a number of standard shapes for the rolling elements such as a ball, cylindrical roller, tapered roller, needle roller, symmetrical and unsymmetrical barrel roller and many more as described by Ocak and Loparo (2004). In this chapter, a ball rolling element is used as was done by Ocak and Loparo (2004).

Three faults are studied in this chapter. These are an inner raceway, an outer raceway, and a rolling element fault. A bearing fault increases the rotational friction of the rotor and, therefore, each fault gives vibration spectra with unique frequency components (Ericsson et al. 2004). It should be taken into account that these

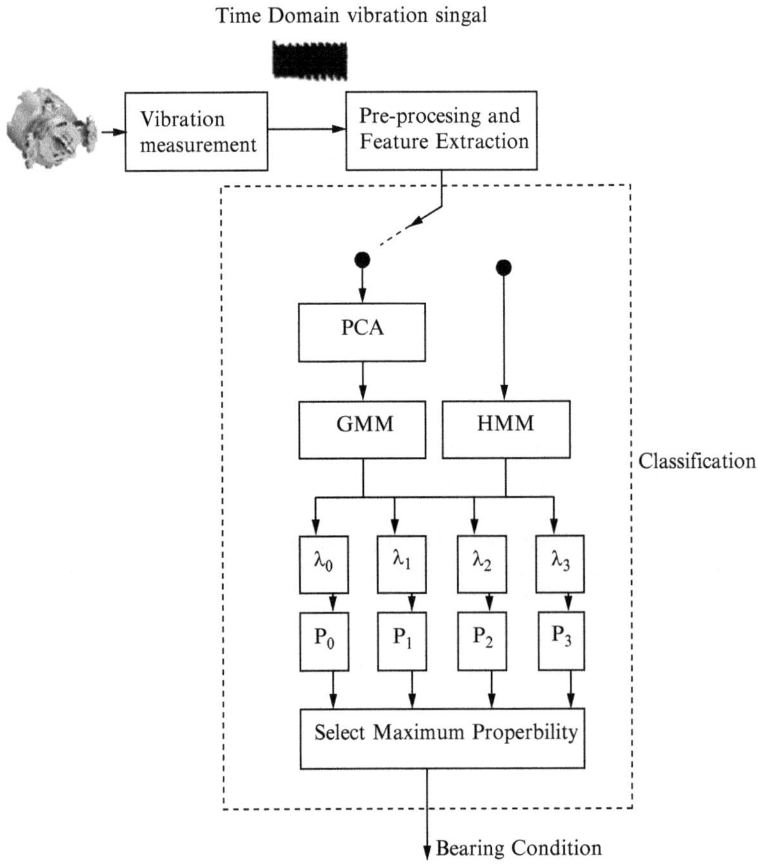

Fig. 6.2 Motor bearing fault detection and diagnosis system

frequency components are a linear function of the running speed and that the two raceway frequencies are also linear functions of the number of balls. The motor bearing condition monitoring systems was implemented by analyzing the vibration signal of all the bearing faults. The vibration signal was produced by Ocak and Loparo (2004) using the impact pulse generated when a ball roller knocks a defect in the raceways or when the defect in the ball knocks the raceways (Li et al. 2000).

The studied motor bearing fault detection and diagnosis system is displayed in Fig. 6.2 (Marwala et al. 2006). The system consists of two major stages after the vibration signal measurement and these are the pre-processing which includes both the feature extraction phase and classification phase.

The initial phase of an automatic fault detection and diagnosis system, as indicated in Fig. 6.3, is signal preprocessing and feature extraction (Marwala et al. 2006). Faults cause a change in the machinery vibration levels and, consequently, the information regarding the health status of the monitored machine is largely contained in the vibration time signal (McClintic et al. 2000). Figure 6.4 shows

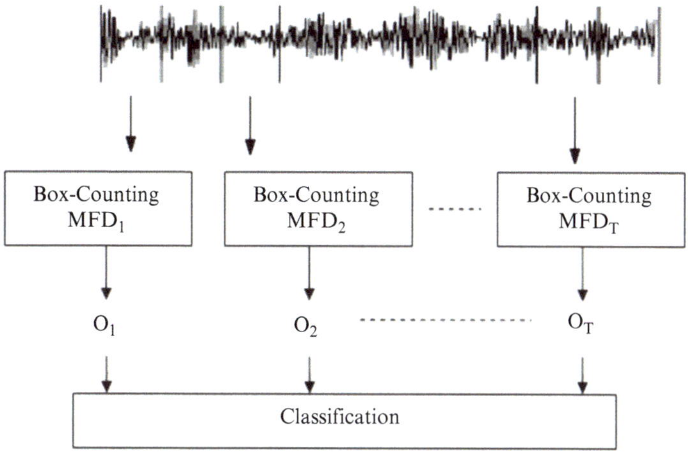

Fig. 6.3 Vibration signal pre-processing and feature extraction

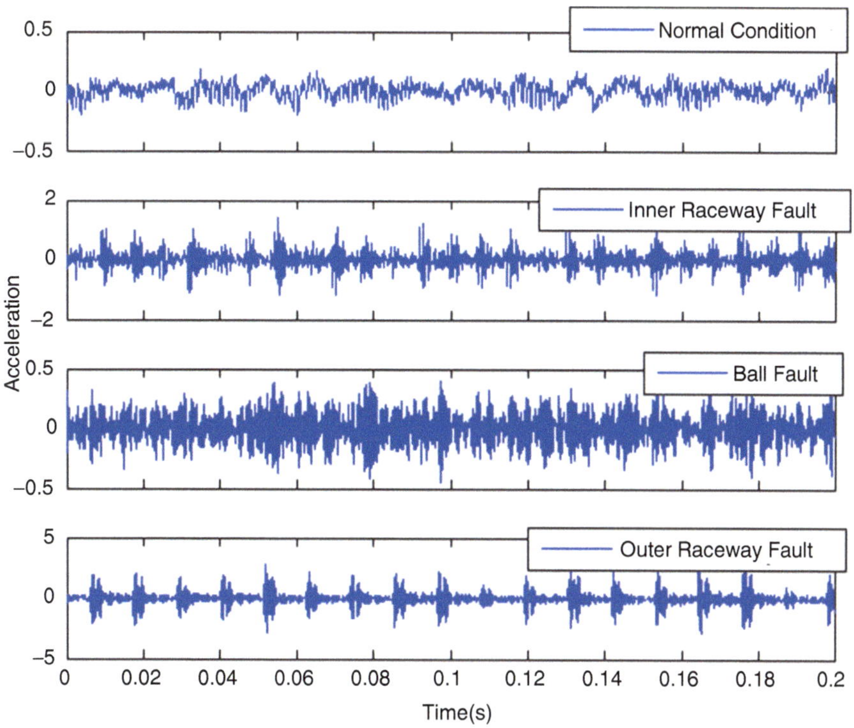

Fig. 6.4 The first 2 s of the vibration signal of the normal, inner raceway fault, ball fault, and outer raceway fault

that the signal is preprocessed by dividing the vibration signals into T windows of equal lengths (Marwala et al. 2006). For this technique to be effective, it should be noted that the width of the window must be more than one revolution of the bearing to ensure that the uniqueness of each vibration fault signal is captured. The preprocessing is followed by extraction of features of each window using the Box-Counting MFD, which forms the observation sequence to be used by the GMM or the HMM classifier. The time domain analysis extracts the non-linear turbulence information of the vibration signal and is expected to give enormous improvement on the performance of the bearing fault detection and diagnosis process.

Due to the large variations of the vibration signal, direct comparison of the signals is difficult. Hence, non-linear pattern classification methods are used to classify different bearing fault conditions. The features extracted were used as inputs to the classification phase of the framework. This chapter compares the performance of the GMM and the HMM classifiers. For the GMM classifier, the principal component analysis (PCA), which was described in Chap. 2, was applied to the feature vector before training to reduce the dimensionality and remove redundant information (Jolliffe 1986). The principal concept behind PCA is to identify the features that explain as much of the total variation in the data as possible with as few of these features as possible. The calculation of the PCA data transformation matrix is based on the eigenvalue decomposition.

The computation of the principal components was conducted as described below (Jolliffe 1986):

- Calculate the covariance matrix of the input data.
- Compute the eigenvalues and eigenvectors of the covariance matrix.
- Preserve the largest eigenvalues and their respective eigenvectors which contains at least 90% of the data.
- Transform the original data into the reduced eigenvectors and, therefore, decrease the number of dimensions of the data.

For more information on the PCA used here to reduce the dimensionality of the feature space, the reader is referred to the work of Jolliffe (1986). In Fig. 6.3, the diagnosis of the motor bearing fault was achieved by calculating the probability of the feature vector, given the entire previously constructed fault model and then the GMM or HMM with maximum probability determined the bearing condition.

This section discusses the experimental database used to evaluate the efficiency of the proposed approach. The performance measure adopted during experimentation is also briefly discussed. The database used to validate the new bearing fault diagnosis discussed in the last section was developed at Rockwell Science Centre by Loparo in 2005. In this data set, single point faults of diameters of 7 mils, 14 mils, and 21 mils (1 mil = 0.001 in.) were introduced using electro-discharge machining. These faults were introduced separately at the inner raceway, rolling element and outer raceway. A more detailed explanation of this data set is presented in (Loparo 2006). The experiments were performed for each fault diameter and this was repeated for two load conditions, which were 1 and 2 hp. The experiment was performed for vibration signals sampled at 12,000 samples per second for the drive

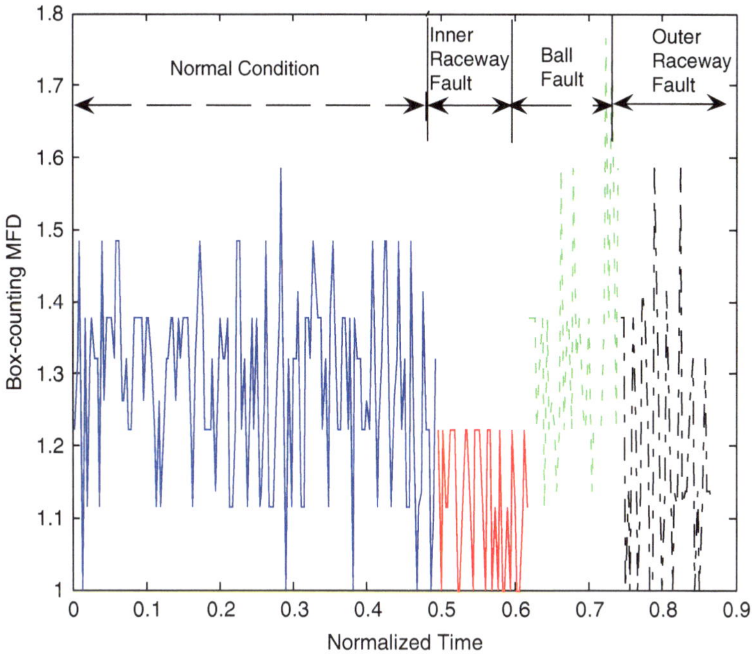

Fig. 6.5 MFD feature extraction comparison for the normal, inner, outer and ball fault for the 1 s vibration

end bearing faults. The vibration signals from this database were divided into equal windows of four revolutions. Half of the resulting sub-signals are used for training and the other half were used for testing.

The main concern was to measure the ability of the system to classify the bearing faults. The performance of the system was measured using the Classification Rate (CR) which is the proportion of fault cases correctly classified.

The optimum HMM architecture, used in the experiment was a 2 state model with a diagonal covariance matrix that contained 10 Gaussian mixtures. The GMM architecture also used a diagonal covariance matrix with three centers. The main advantage of using the diagonal covariance matrix in both cases was that this de-correlated the feature vectors. This was necessary because fractal dimensions of adjacent scales were highly correlated (Maragos and Potamianos 1999).

The first set of experiments measured the effectiveness of the time-domain fractal dimension based feature-extraction using vibration signal of the faults as shown in Fig. 6.5 (Marwala et al. 2006).

Figure 6.5 shows the first 2 s of the vibration signals used. It can be clearly seen that there is fault specific information which must be extracted. Figure 6.6 shows the MFD feature vector which extracts the bearing's fault specific information (Marwala et al. 2006). It should be noted that these features are only for the first second of the vibration signal. Figure 6.6 clearly shows that the presented feature extraction

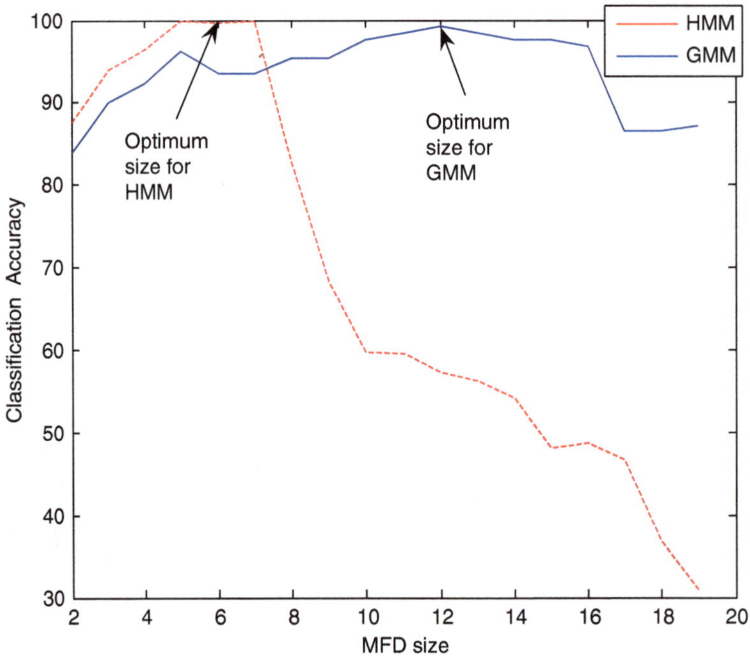

Fig. 6.6 The graph of the change classification rate with change in MFD size

Table 6.1 The classification rate for different loads and fault diameters for the GMM and HMM classifier

Load	7 mils	14 mils	21 mils	Load	7 mils	14 mils
	HMM	GMM	HMM	GMM	HMM	GMM
1	100%	99.2%	100%	98.7%	100%	99%
2	100%	99%	100%	99.1%	100%	99%

method does indeed extract the fault specific features which are used to classify different bearing faults (Marwala et al. 2006). For this reason, the presented MFD feature extraction is expected to give enormous improvement to the performance of the bearing fault detection and diagnosis. Nevertheless, the optimum size of the MFD must be initially found. Figure 6.6 shows the graph of change of the system accuracy with the change of the MFD size. The figure shows that the GMM generally has a large optimum MFD size of 12 compared to 6 for HMM.

Having used the optimum HMM and GMM architecture discussed previously, the classification accuracy that was found for different bearing loads and different bearing fault diameters appears in Table 6.1 for the GMM and the HMM classifier.

Table 6.1 shows that the HMM outperforms the GMM classifier for all cases, with a 100% and 99.2% classification rate for HMM and GMM, respectively. Table 6.1 also shows that changing the bearing load or diameter does not significantly change the classification rate.

Using a Pentium IV with 2.4 GHz processor speed, further experimenting showed that the average training time of HMM was 19.5 s. This was more than 20 times higher than the GMM training time, which was found to be 0.83 s. In summary, even though HMM gave higher classification rate when compared to GMM it was time consuming to train the models when compared to GMM. It is probably worth mentioning that, it was observed that using the PCA dimension reduction technique does not affect the classification rate. Nevertheless, this reduced the dimension from 84 to 11, which makes GMM training even more computationally efficient when compared to training the HMM.

This chapter presented the obtained using MFD short time feature extraction. The results demonstrated that this technique does extract fault specific features. Furthermore, the results showed that for the GMM classifier using PCA, the classification rate was not affected; it simply reduced the dimensionality of the input feature vector which makes the GMM models less complex than the HMM models. Further experimentation revealed that there was an optimum MFD size which gave the optimum classification rate. From the results obtained, it was found that the GMM generally had larger optimum MFD size than the HMM.

The second set of tests that were performed compared the performance of GMM and HMM in classifying the different bearing faults. The test revealed that the HMM outperformed the GMM classifier with a classification rate of 100%. Further testing of these classifiers revealed that, the major disadvantage of the HMM classifier was that it took longer to train than the GMM classifier, even though GMM had larger MFD size than HMM. So, it is recommended that one use the GMM classifier when time is the major issue in that particular application. It was further observed that changing the bearing load or diameter does not significantly affect the classification rate of the presented framework.

6.4 Conclusions

A framework that used a time-domain fractal-based feature extraction method to extract the non-linear turbulent information of the vibration signal has been presented. Using these features together with HMM and GMM classifiers, the results showed that the HMM classifier outperformed the GMM classifier with the HMM giving 100% and the GMM 99.2% classification rate. Nevertheless, the major drawback of the HMM classifier was that it was computationally expensive, taking 20 times longer than the GMM classifier to train.

References

Batty M (1985) Fractals – geometry between dimensions. New Sci 105:31–35

Baum L (1972) An inequality and associated maximization techniques in statistical estimation for probabilistic function of Markov processes. Inequality 3:1–8

Bordes L, Chauveau D, Vandekerkhove (2007) A stochastic EM algorithm for a semiparametric mixture model. J Comput Stat Data Anal 51:5429–5443
Boutros T, Liang M (2011) Detection and diagnosis of bearing and cutting tool faults using hidden Markov models. Mech Syst Signal Process 25:2102–2124
Briggs J (1992) Fractals: the patterns of chaos. Thames and Hudson, London
Caelli T, McCabe A, Briscoe G (2001) Shape tracking and production using hidden Markov models. In: Bunke H, Caelli T (eds) Hidden Markov models: applications in computer vision. World Scientific, Singapore
Calefati P, Amico B, Lacasella A, Muraca E, Zuo MJ (2006) Machinery faults detection and forecasting using hidden Markov models. In: Proceedings of the 8th biennial ASME conference on engineering systems design and analysis, Torino, Italy, pp 895–901
Ching WK, Ng M (2006) Markov chains: models, algorithms and applications, International series on operations research and management science. Springer, New York
Ching WK, Ng M, Fung E (2003) Higher-order hidden Markov models with applications to DNA sequences. In: Liu J, Cheung Y, Yin H (eds) Intelligent data engineering and automated learning. Springer, New York
Ching WK, Ng M, Wong K (2004) Hidden Markov model and its applications in customer relationship management. IMA J Manag Math 15:13–24
Dempster AP, Laird NM, Rubin DB (1977) Maximum likelihood from incomplete data via the EM algorithm. J Roy Stat Soc B 39:1–38
Ericsson S, Grip N, Johansson E, Persson LE, Sjöberg R, Strömberg JO (2004) Towards automatic detection of local bearing defects in rotating machines. Mech Syst Signal Process 19:509–535
Falconer K (2003) Fractal geometry: mathematical foundations and applications. Wiley, New Jersey
Ikizoglu S, Caglar R, Seker S (2010) Hurst parameter and fractal dimension concept for fault detection in electric motors. Int Rev Electric Eng 5:980–984
Ingrassia S, Rocci R (2007) A stochastic EM algorithm for a semiparametric mixture model. Comput Stat Data Anal 51:5429–5443
Jolliffe IT (1986) Principal component analysis. Springer, New York
Kang S, Wang B, Kang Y (2006) Early detection for short-circuit fault in low-voltage systems based on fractal exponent wavelet analysis. In: Proceedings of the SPIE – the international society for optical engineering: San Diego, California, art. no. 63574Z
Kauermann G, Xu R, Vaida F (2007) Nonlinear random effects mixture models: maximum likelihood estimation via the EM algorithm. Comput Stat Data Anal 51:6614–6623
Koski T (2001) Hidden Markov models for bioinformatics. Kluwer Academic, Dordrecht
Lee S, Li L, Ni J (2010) Online degradation assessment and adaptive fault detection using modified hidden Markov model. J Manufacturing Sci Eng Trans ASME 132:0210101–02101011
Li B, Chow MY, Tipsuwan Y, Hung JC (2000) Neural-network-based motor rolling bearing fault diagnosis. IEEE Trans Ind Electron 47:1060–1068
Li DH, Wang JF, Shi LT (2005) Application of fractal theory in DC system grounding fault detection. Automation Electric Power Syst 29:53–56+84
Loparo KA (2006) Bearing data center seeded fault test data. http://www.eecs.case.edu/-laboratory/bearing/download.htm. Last accessed 01 June 2006
Lou X, Loparo KA, Discenzo FM, Yoo J, Twarowski A (2004) A model-based technique for rolling element bearing fault detection. Mech Syst Signal Process 18:1077–1095
Ma J (2009) The application of fractal analysis to fault detection and diagnoses in course of welded. In: Proceedings of the 2nd international conference on model and sim, Manchester, UK, pp 263–266
Mandelbrot BB (1982) The fractal geometry of nature. W.H. Freeman and Co, San Francisco
Maragos P, Potamianos A (1999) Fractal dimensions of speech sounds: computation and application to automatic speech recognition. J Acoust Soc Am 105:1925–1932
Maragos P, Sun FK (1993) Measuring the fractal dimension of signals: morphological covers and iterative optimization. IEEE Trans Signal Process 41:108–121

Marwala T, Mahola U, Nelwamondo FV (2006) Hidden Markov models and Gaussian mixture models for bearing fault detection using fractals. In: Proceedings of the IEEE international conference on neural networks, Vancouver, Canada, pp 3237–3242

McClintic K, Lebold M, Maynard K, Byington C, Campbell R (2000) Residual and difference feature analysis with transitional gearbox data. In: Proceedings of the 54th meeting of the society for machinery failure prevention technology, Virginia Beach, pp 635–645

McFadden PD, Smith JD (1984) Vibration monitoring of rolling element bearings by high frequency resonance technique – a review. Tribol Int 77:3–10

McLachlan G, Krishnan T (1997) The EM algorithm and extensions, Wiley series in probability and statistics. Wiley, New Jersey

Menon S, Kim K, Uluyol O, Nwadiogbu EO (2003) Incipient fault detection and diagnosis in turbine engines using hidden Markov models. Am Soc Mech Eng Int Gas Turbine Inst, Turbo Expo (Publication) IGTI:493–500

Ocak H, Loparo KA (2004) Estimation of the running speed and bearing defect frequencies of an induction motor from vibration data. Mech Syst Signal Process 18:515–533

Park J, Qian GQ, Jun Y (2007) Using the revised EM algorithm to remove noisy data for improving the one-against-the-rest method in binary text classification. Info Process Manag 43:1281–1293

Patel AK, Patwardhan AW, Thorat BN (2007) Acceleration schemes with application to the EM algorithm. Comput Stat Data Anal 51:3689–3702

Pickover CA (2009) The math book: from Pythagoras to the 57th dimension, 250 milestones in the history of mathematics. Sterling Publishing Company, New York

Purushothama V, Narayanana S, Suryana-rayana AN, Prasad B (2005) Multi-fault diagnosis of rolling bearing elements using wavelet analysis and hidden Markov model based fault recognition. NDT&E Int 38:654–664

Rabiner LR (1989) A tutorial on hidden Markov models and selected applications in speech recognition. Proc IEEE 77:257–286

Reynolds DA (1992) A Gaussian mixture modeling approach to text-independent speaker identification. PhD thesis, Georgia Institute of Technology

Reynolds DA, Rose RC (1995) Robust text-independent speaker identification using Gaussian mixture speaker models. IEEE Trans Acoust Speech Signal Process 3:72–83

Reynolds DA, Quatieri TF, Dunn RB (2000) Speaker verification using adapted Gaussian mixture models. Digit Signal Process 10:19–41

Russ JC (1994) Fractal surfaces. Springer, New York

Shanlin K, Baoshe L, Feng F, Songhua S (2007) Vibration fault detection and diagnosis method of power system generator based on wavelet fractal network. In: Proceedings of the 26th Chinese control conference, Zhangjiajie, China, pp 520–524

Smyth P (1994) Hidden Markov models for fault detection in dynamic systems. Pattern Recognit 27:149–164

Song Y, Wang G, Chen X (2007) Fault detection and analysis of distributed power system short-circuit using wavelet fractal network. In: Proceedings of the 8th international conference on electron measurement and instruments, Xi'an, China, pp 3422–3425

Sornette D (2004) Critical phenomena in natural sciences: chaos, fractals, selforganization, and disorder: concepts and tools. Springer, New York

Tai AH, Ching WK, Chan LY (2009) Detection of machine failure: hidden Markov model approach. Comput Ind Eng 57:608–619

Viterbi A (1967) Error bounds for convolutional codes and an asymptotically optimum decoding algorithm. IEEE Trans Info Theory 13:260–269

Wang F, Zheng F, Wu W (2000) A C/V segmentation method for Mandarin speech based on multi-scale fractal dimension. In: Proceedings of the international conference on spoken language process, Beijing, China, pp 648–651

Wang H, Hu Z (2007) Speeding up HMRF_EM algorithms for fast unsupervised image segmentation by bootstrap resampling: application to the brain tissue segmentation. Signal Process 87:2544–2559

Wang YF, Kootsookos PJ (1998) Modeling of low shaft speed bearing faults for condition monitoring. Mech Syst Signal Process 12:415–426

Wang X, Schumitzky A, D'Argenio DZ (2007) Constrained monotone EM algorithms for finite mixture of multivariate Gaussians. Comput Stat Data Anal 51:5339–5351

Wong WC, Lee JH (2010) Fault detection and diagnosis using hidden Markov disturbance models. Ind Eng Chem Res 49:7901–7908

Yan H, Zhang XC, Li G, Yin J, Cheng W (2007) Fault detection for wall-climbing robot using complex wavelet packets transform and fractal theory. Acta Photonica Sinica 36:322–325

Yang XB, Jin XQ, Du ZM, Zhu YH (2011) A novel model-based fault detection method for temperature sensor using fractal correlation dimension. Build Environ 46:970–979

Zhang HT, An Q, Hu ZK, Chen ZW (2010) Fault detection wavelet fractal method of circuit of three-phase bridge rectifier. In: Proceedings of the international conference on intelligent systems design and engineering applications, Cairo, Egypt, pp 725–729

Zhou SY, Wang SQ (2005) On-line fault detection and diagnosis in industrial processes using hidden Markov model. Dev Chem Eng Miner Process 13:397–406

Chapter 7
Fuzzy Systems for Condition Monitoring

7.1 Introduction

Fuzzy set theory (FST) has been successfully applied in a number of applications in the past (Dhlamini and Marwala 2005; Dhlamini et al. 2006; Dhlamini 2007). Successful applications of FST include relating operators to chemical plants based on their skill availability, health and age (Majozi and Zhu 2005), in control systems (Kubica et al. 1995), in pattern recognition (Flaig et al. 2000), applied FST in pattern recognition and in the evaluation of state government performance (Ammar and Wright 2000). The principal advantage of the FST is its capability to model uncertain data that many systems and environments exhibit. FST enables the exploration of the interaction of variables which define a system, and how the variables affect the system's output.

This chapter applies FST for the condition monitoring of transformer bushings. In transformer bushings the IEC60599 standard can be applied to evaluate the range associated with normal, elevated and abnormal concentrations of gas. For this chapter the FST was applied to evaluate the extent of how high is too high (or too low) so that the elevated (or depressed) condition must be classified as dangerous and require the transformer bushing to be decommissioned from service.

In essence, this chapter presents the application of FST to diagnose the condition of high voltage bushings. This procedure applies Dissolved Gas Analysis (DGA) data from bushings based on the IEC60599, California State University Sacramento (CSUS) and IEEE C57-104 criteria for Oil Impregnated Paper (OIP) bushings. FST was applied to evaluate the interrelations that exist between each bushing's identifying attributes, i.e., the dissolved gases in oil. In DGA there is a relationship between the resulting failure and the concurrent existence of oxygen with other gases, for instance, hydrogen, methane, ethane, ethylene, acetylene, and carbon monoxide in a bushing. The incidence of combustible gases and the nonexistence of oxygen is itself not a sign of imminent failure. Implementing FST on bushing data is essential because the degree to which the assessment standard is under the cut-off point for a safe operation is not uniform for each bushing. This inconsistency

T. Marwala, *Condition Monitoring Using Computational Intelligence Methods: Applications in Mechanical and Electrical Systems*, DOI 10.1007/978-1-4471-2380-4_7,

can be accounted for in the assessment procedure by applying fuzzy set theory. Temperature is a vital measure in the evaluation process and denotes both to the operating temperature of the oil and the difference between the ambient and the oil temperature.

Bushings that constantly function at temperatures near or above the auto-ignition temperature of any of the gases or oil have a considerably higher probability of explosion than those that operate at lower temperatures with the same ratio of gases (Dhlamini et al. 2006). The *auto-ignition temperature* of a substance is the temperature at or above which a material will unexpectedly catch fire in the absence of an external spark or flame.

Msiza et al. (2011) applied neural networks to the detection of transformer faults and in particular for evaluating the relevance of the input space parameters. They applied a multi-layer neural network, initially populated with all the ten input parameters (10 V-Model). A matrix containing causal information about the possible relevance of each input parameter was then obtained. The information from this matrix was proven to be valid through the construction and testing of another two, separate, multilayer networks. One network's input space was populated with the five most relevant parameters (MRV-Model), while the other was populated with the five least relevant parameters (LRV-Model). The obtained classification accuracy values were as follows: 100% for the 10 V-Model, 98.5% for the MRV-Model, and 53.0% for the LRV-Model.

Auto-ignition temperature must not be mistaken for flash or fire points, which are normally a couple of hundred degrees lower (Dhlamini et al. 2006). The flash point is the lowest temperature at which a liquid can create an ignitable mixture with air near the surface of the liquid. In this regard, the lower the flash point, the easier it is to ignite the material. *Fire point* is the minimum sample temperature at which vapor is formed at an adequate rate to sustain combustion and it is the lowest temperature at which the ignited vapor continues to burn for at least 5 s (Dhlamini et al. 2006).

Flash point is identified by the ASTM D 93 technique known as the "Flash Point by Pensky-Martens Closed Tester" for fuel oils (Dhlamini et al. 2006) or otherwise by the ASTM D 92 method known as the "Flash and Fire Points by Cleveland Open Cup". At the fire point, the temperature of the flame goes into self-sustainability in order to continue burning the liquid, whereas at the flash point the flame does not require to be sustained. The fire point is typically a few degrees above the flash point.

In this chapter, ten identifying attributes were chosen to develop membership functions and these were the concentrations of hydrogen, oxygen, nitrogen, methane, carbon monoxide, carbon dioxide, ethylene, ethane, acetylene and total dissolved combustibles gases.

Boesack et al. (2010) applied genetic algorithm and fuzzy logic for automatic generation control to rationalize the fuzzy inference rules and the appropriate selection of the input and output membership functions. They fed the fuzzy logic controller with certain parameters which can be optimized using a genetic algorithm to suit the specific application under control.

Sainz Palmero et al. (2005) applied fuzzy ARTMAP for the detection of faults in Alternating Current motors. When the system was tested a good level of detection and classification was obtained. Furthermore, the knowledge extracted had an acceptable degree of interpretability. Korbicz and Kowal (2007) applied fuzzy systems to the fault detection of valves in the industrial installation of the Lublin sugar factory. The results indicated the effectiveness of the technique. Mendonca et al. (2009) applied a fuzzy system for the detection of faults and isolation of an industrial valve and could detect and isolate the simulated abrupt and incipient faults.

Razavi-Far et al. (2009) applied fuzzy systems for fault detection and isolation in a steam generator while elsewhere D'Angelo et al. (2011) applied fuzzy systems for the detection of incipient faults in induction machine stator-windings. Chen et al. (2008) applied fuzzy system for fault detection in railway track circuits and elsewhere, Evsukoff and Gentil (2005) applied fuzzy system for fault detection and isolation in nuclear reactors. Other applications of fuzzy systems are in fault detection of analogue circuits (Catelani et al. 2002), for condition monitoring of a packaging plant (Jeffries et al. 2001), for tool wear condition monitoring (Aliustaoglu et al. 2009), for monitoring water pipelines (Lau and Dwight 2011) and for machine condition monitoring (Javadpour and Knapp 2003a).

7.2 Computational Intelligence

Computational Intelligence has many tools in its toolbox. This section explains the use of six of these tools: Basic Fuzzy Logic Theory, Membership Functions, Fuzzy Rules, Decisions Based on Fuzzy Rules, Aggregated Rules and Defuzzification.

7.2.1 Fuzzy Logic Theory

Fuzzy logic is a method of mapping an input space to an output space by means of a list of linguistic rules that entail the *if-then* statements (Bih 2006; Marwala and Lagazio 2011). Basic fuzzy logic is made up of four components: fuzzy sets, membership functions, fuzzy logic operators, and fuzzy rules (Von Altrock 1995; Biacino and Gerla 2002; Cox 1994; Marwala and Lagazio 2011).

In classical set theory, an object is either an element or is not an element of a given set (Devlin 1993; Ferreirós 1999; Johnson 1972; Marwala and Lagazio 2011). Accordingly, it is conceivable to define if an object is an element of a given set since a set has distinctive boundaries, providing that such an object cannot take on fractional membership. An alternate method of seeing this is that an object's belonging to a set is either true or false. A characteristic function for a classical set has a value of one if the object is an element of the set and a value of zero if the object is not an element of a set (Cantor 1874). For example, if a set X is defined to characterize the possible heights of all people, one could define a "tall" subset for any person who is above or equal to a specific height x, and anyone below x doesn't belong to the "tall" subset but belongs to a "short" subset. This is obviously inflexible as a person just below the boundary is categorized as being short when they are clearly "almost tall". Here, vague values such as "reasonably tall" are not permitted. Furthermore, such clear-cut defined boundaries can be very subjective in terms of what different people may define as belonging to a specific set.

The crucial aim of fuzzy logic is to allow a more flexible representation of sets of objects by applying fuzzy sets. A fuzzy set does not have the perfect margins as does a classical set; the objects are characterized by a *degree* of membership to a specific set (Hájek 1995; Halpern 2003; Wright and Marwala 2006; Hájek 1998). Accordingly, transitional values of objects can be characterized in a way that is nearer to the way that the human brain thinks, compared to the clear cut-off margins in classical sets.

A *membership function* expresses the *degree* that an object is an element of a certain set or class. The membership function is a curve that maps the input space variable to a number between 0 and 1, signifying the degree that a specific input variable is an element of a specific set (Klir and Folger 1988; Klir et al. 1997; Klir and Yuan 1995). A membership function can be a curve of any shape. Expanding the example above, there are two subsets: one for tall people and one for short that overlap. In this way, a person can have a partial participation in each of these sets, consequently determining the degree to which the person is both tall and short.

Logical operators are defined to produce new fuzzy sets from the existing fuzzy sets. In classical set theory, there are three key operators used, permitting logical expressions to be defined: intersection, union, and the complement (Kosko 1993; Kosko and Isaka 1993). These operators are also used in fuzzy logic, but have been modified to deal with partial memberships. The intersection (AND operator) of two fuzzy sets is given by a minimum operation, and the union (OR operator) of two fuzzy sets is given by a maximum operation (Novák 1989, 2005; Novák et al. 1999). These logical operators are used in the rules and determination of the final output fuzzy set.

Fuzzy rules express the conditional statements which are used to model the input–output relationships of the system, and are articulated in natural language. These linguistic rules are in the form of *if-then* statements which use the logical operators and membership functions to produce an output. A vital characteristic of fuzzy logic is the use of linguistic variables. Linguistic variables are variables that use words or sentences as their values as an alternative to numbers (Zadeh 1965; Zemankova-Leech 1983; Zimmermann 2001; Marwala and Lagazio 2011). Each linguistic variable takes on a linguistic value that corresponds to a fuzzy set. The set of values that it can take on is called the *term set*. For instance, a linguistic variable *height* could have the following term set {*very tall, tall, medium, short, very short*}. A single fuzzy *if-then* rule assumes the form if x is A then y is B.

Here A and B are linguistic values defined by fuzzy sets on the ranges (universes of discourse) X and Y, respectively.

Every one of the rules is assessed for an input set, and the corresponding output for the rule is attained. If an input corresponds to two linguistic variable values then the rules associated with both these values will be assessed. Moreover, the rest of the rules will be assessed; nevertheless they will not have an influence on the final result as the linguistic variable will have a value of zero. Consequently, if the antecedent is true to some degree, the result will have to be true to some degree (Zadeh 1965). The degree of each linguistic output value is then calculated by performing a logical sum for each membership function (Zadeh 1965), after which all the sums for a specific linguistic variable can be combined. These last phases comprise the use of an inference technique which will map the result onto an output membership function (Zadeh 1965).

Lastly the *defuzzification* process is accomplished where a single numeric output is produced. One technique for computing the degree of each linguistic output value is to take the maximum of all rules describing this linguistic output value, the output is taken as the center of gravity of the area under the affected part of the output membership function. There are other inference techniques such as averaging and sum mean square. Figure 7.1 displays the stages involved in generating input–output mapping using fuzzy logic.

The application of series of fuzzy rules and inference approaches to yields defuzzified output constitute a Fuzzy Inference System (FIS). The final manner in which the aggregation process takes place and the technique of defuzzification differs, depending on the application of the selected FIS. The method explained below is that of Mamdani (1974).

There are a number of kinds of fuzzy inference systems which differ according to the fuzzy reasoning and the form of the *if-then* statements applied. One of these approaches is the Takagi-Sugeno-Kang neuro-fuzzy technique (Takagi and Sugeno 1985; Araujo 2008). This method is similar to the Mamdani method described above, except that the consequent part is of a different form and, as a result, the defuzzification technique is different. The *if-then* statement of a Sugeno fuzzy system expresses the output of each rule as a function of the input variables (Sugeno and Kang 1988; Sugeno 1985; Takagi and Sugeno 1985; Babuska 1991). Applications of neuro-fuzzy model include its use in modeling liquid-holdup prediction in horizontal Multiphase flows (El-Sebakhy 2010), in reinforcement group cooperation-based symbiotic evolution (Hsu and Lin 2009), in gene extraction for cancer diagnosis (Huang and Kecman 2005), in control of nonlinear system (Iplikci 2010), for autonomous parallel parking (Demirli and Khoshnejad 2009), for conflict modeling (Tettey and Marwala 2006), and in constitutive modeling (Cabalar et al. 2010).

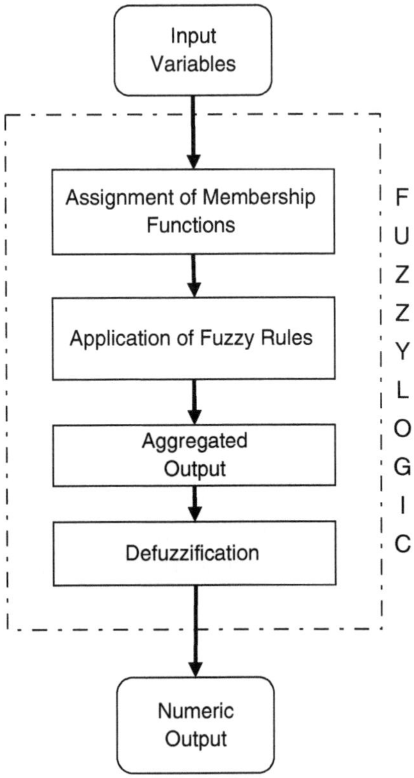

Fig. 7.1 Showing the steps involved in the application of fuzzy logic to a problem (Wright and Marwala 2006)

7.2.2 Membership Functions

Membership functions are the most significant stage in fuzzy set theory applications and this step takes the most time and should be precise. Membership function curves that can be used include straight line, the Gaussian function, the sigmoid, and polynomial functions. Bojadziev and Bojadziev (1995) discussed that triangular functions accurately represented most memberships. In general, triangular and trapezoidal membership functions are representative of most cases (Majozi and Zhu 2005; Zadeh 1973). For this chapter the trapezoidal and triangular membership functions were chosen to model the safe operating limits for gas contaminants inside the bushing's oil and each of the attributes was rated in as being high, medium or low (Dhlamini 2007). The rating depends on the measured magnitude of the attribute compared to the reject threshold obtained in the IEC60599 criteria. The membership functions were concentrations of hydrogen, oxygen, nitrogen, methane, carbon monoxide, carbon dioxide, ethylene, ethane, acetylene and total dissolved combustibles gases as given in Eqs. 7.1–7.30 as described by Dhlamini (2007) and Dhlamini et al. (2006).

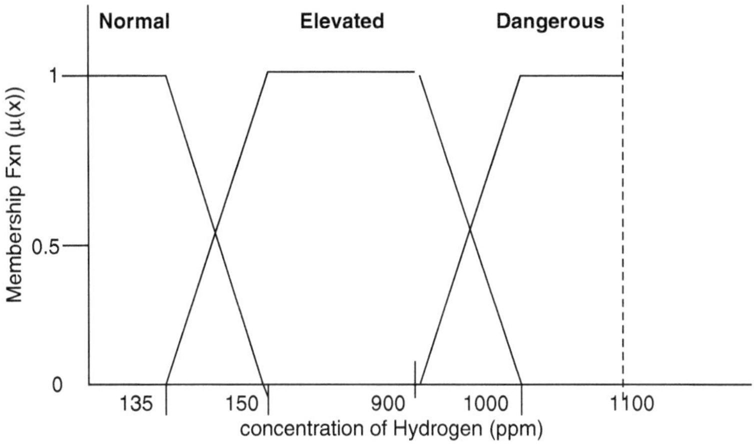

Fig. 7.2 Membership functions of Hydrogen

Hydrogen

$$\mu_{Normal}(x) = \begin{cases} 1 & \text{for } 0 \leq x \leq 135 \\ -0.067x + 10 & \text{for } 135 \leq x \leq 150 \end{cases} \tag{7.1}$$

$$\mu_{\text{Elevated}}(x) = \begin{cases} 0.067x - 9 & \text{for } 135 \leq x \leq 150 \\ 1 & \text{for } 150 \leq x \leq 900 \\ -0.067x + 10 & \text{for } 900 \leq x \leq 1000 \end{cases} \tag{7.2}$$

$$\mu_{\text{Dangerous}}(x) = \begin{cases} 0.01x - 9 & \text{for } 900 \leq x \leq 1000 \\ 1 & \text{for } x \geqslant 1000 \end{cases} \tag{7.3}$$

The membership function for Eqs. 7.1–7.3 is shown in Fig. 7.2 and Eqs. 7.4–7.30 have similar membership functions.

Methane

$$\mu_{Normal}(x) = \begin{cases} 1 & \text{for } 0 \leq x \leq 23 \\ -0.5x + 12.5 & \text{for } 23 \leq x \leq 25 \end{cases} \tag{7.4}$$

$$\mu_{\text{Elevated}}(x) = \begin{cases} 0.5x - 11.5 & \text{for } 23 \leq x \leq 25 \\ 1 & \text{for } 25 \leq x \leq 72 \\ -0.125x + 10 & \text{for } 72 \leq x \leq 80 \end{cases} \tag{7.5}$$

$$\mu_{\text{Dangerous}}(x) = \begin{cases} 0.125x - 9 & \text{for } 72 \leq x \leq 80 \\ 1 & \text{for } x \geq 80 \end{cases} \tag{7.6}$$

Ethane

$$\mu_{Normal}(x) = \begin{cases} 1 & \text{for } 0 \le x \le 9 \\ -x + 10 & \text{for } 9 \le x \le 10 \end{cases} \tag{7.7}$$

$$\mu_{\text{Elevated}}(x) = \begin{cases} x - 9 & \text{for } 9 \le x \le 10 \\ 1 & \text{for } 10 \le x \le 32 \\ -0.333x + 11.66 & \text{for } 32 \le x \le 35 \end{cases} \tag{7.8}$$

$$\mu_{\text{Dangerous}}(x) = \begin{cases} 0.333x - 10.66 & \text{for } 32 \le x \le 35 \\ 1 & \text{for } x \ge 35 \end{cases} \tag{7.9}$$

Ethylene

$$\mu_{Normal}(x) = \begin{cases} 1 & \text{for } 0 \le x \le 18 \\ -0.5x + 10 & \text{for } 18 \le x \le 20 \end{cases} \tag{7.10}$$

$$\mu_{\text{Elevated}}(x) = \begin{cases} 0.5x - 9 & \text{for } 18 \le x \le 20 \\ 1 & \text{for } 20 \le x \le 90 \\ -0.1x + 10 & \text{for } 90 \le x \le 100 \end{cases} \tag{7.11}$$

$$\mu_{\text{Dangerous}}(x) = \begin{cases} 0.1x - 9 & \text{for } 90 \le x \le 100 \\ 1 & \text{for } x \ge 100 \end{cases} \tag{7.12}$$

Acetylene

$$\mu_{Normal}(x) = \begin{cases} 1 & \text{for } 0 \le x \le 14 \\ -x + 15 & \text{for } 14 \le x \le 15 \end{cases} \tag{7.13}$$

$$\mu_{\text{Elevated}}(x) = \begin{cases} x - 14 & \text{for } 14 \le x \le 15 \\ 1 & \text{for } 15 \le x \le 63 \\ -0.142857x + 10 & \text{for } 63 \le x \le 70 \end{cases} \tag{7.14}$$

$$\mu_{\text{Dangerous}}(x) = \begin{cases} 0.142857x - 9 & \text{for } 63 \le x \le 70 \\ 1 & \text{for } x \ge 70 \end{cases} \tag{7.15}$$

Carbon Monoxide

$$\mu_{Normal}(x) = \begin{cases} 1 & \text{for } 0 \le x \le 450 \\ -0.02x + 10 & \text{for } 450 \le x \le 500 \end{cases} \tag{7.16}$$

$$\mu_{\text{Elevated}}(x) = \begin{cases} 0.02x - 9 & \text{for } 450 \le x \le 500 \\ 1 & \text{for } 500 \le x \le 900 \\ -0.01x + 10 & \text{for } 900 \le x \le 1000 \end{cases} \tag{7.17}$$

$$\mu_{\text{Dangerous}}(x) = \begin{cases} 0.01x - 9 & \text{for } 900 \leq x \leq 1000 \\ 1 & \text{for } x \geq 1000 \end{cases} \tag{7.18}$$

Nitrogen

$$\mu_{Normal}(x) = \begin{cases} 1 & \text{for } 0 \leq x \leq 0.9 \\ -10x + 10 & \text{for } 0.9 \leq x \leq 1 \end{cases} \tag{7.19}$$

$$\mu_{\text{Elevated}}(x) = \begin{cases} 10x - 9 & \text{for } 0.9 \leq x \leq 1 \\ 1 & \text{for } 1 \leq x \leq 9 \\ -x + 10 & \text{for } 9 \leq x \leq 10 \end{cases} \tag{7.20}$$

$$\mu_{\text{Dangerous}}(x) = \begin{cases} x - 9 & \text{for } 9 \leq x \leq 10 \\ 1 & \text{for} x \geq 10 \end{cases} \tag{7.21}$$

Oxygen

$$\mu_{Normal}(x) = \begin{cases} 1 & \text{for } 0 \leq x \leq 0.09 \\ -100x + 10 & \text{for } 0.09 \leq x \leq 0.1 \end{cases} \tag{7.22}$$

$$\mu_{\text{Elevated}}(x) = \begin{cases} 100x - 9 & \text{for } 0.09 \leq x \leq 0.10 \\ 1 & \text{for } 0.10 \leq x \leq 0.18 \\ -50x + 10 & \text{for } 0.18 \leq x \leq 0.20 \end{cases} \tag{7.23}$$

$$\mu_{\text{Dangerous}}(x) = \begin{cases} 50x - 9 & \text{for } 0.18 \leq x \leq 0.20 \\ 1 & \text{for } x \geq 0.20 \end{cases} \tag{7.24}$$

Carbon Dioxide

$$\mu_{Normal}(x) = \begin{cases} 1 & \text{for } 0 \leq x \leq 9000 \\ -0.001x + 10 & \text{for } 9000 \leq x \leq 10000 \end{cases} \tag{7.25}$$

$$\mu_{\text{Elevated}}(x) = \begin{cases} 0.001x - 9 & \text{for } 9000 \leq x \leq 10000 \\ 1 & \text{for } 10000 \leq x \leq 13500 \\ -0.00067x + 10 & \text{for } 13500 \leq x \leq 15000 \end{cases} \tag{7.26}$$

$$\mu_{\text{Dangerous}}(x) = \begin{cases} 0.00067x - 9 & \text{for } 13500 \leq x \leq 15000 \\ 1 & \text{for } x \geq 15000 \end{cases} \tag{7.27}$$

Total Combustible Gases

$$\mu_{Normal}(x) = \begin{cases} 1 & \text{for } 0 \leq x \leq 648 \\ -0.01389x + 10 & \text{for } 648 \leq x \leq 720 \end{cases} \tag{7.28}$$

$$\mu_{\text{Elevated}}(x) = \begin{cases} 0.01389x - 9 & \text{for } 648 \le x \le 720 \\ 1 & \text{for } 720 \le x \le 4500 \\ -0.002x + 10 & \text{for } 4500 \le x \le 5000 \end{cases} \tag{7.29}$$

$$\mu_{\text{Dangerous}}(x) = \begin{cases} 0.002x - 9 & \text{for } 4500 \le x \le 5000 \\ 1 & \text{for } x \ge 5000 \end{cases} \tag{7.30}$$

7.2.3 *Fuzzy Rules*

Fuzzy rules represent the interrelation between all the inputs. This is a point where user experience can be integrated in the mathematical modeling. It has been estimated by many researchers before, that the number of rules is theoretically equal to the number of fuzzy categories raised to the power of the number of fuzzy criteria. Fuzzy categories used here were: the membership functions "dangerous", "elevated" or "normal". Fuzzy criteria applied here were the different gases that were present, *i.e.*, hydrogen, methane, ethane, ethylene, acetylene, carbon monoxide, nitrogen, oxygen, carbon dioxide and total combustible gases. The rates of change of the gases were not used because the available data was taken on 1 day only.

Rules have an antecedent and a consequence. Rules can be expressed in the form (Ammar and Wright 2000; Wang 2000; Mamdani and Assilian 1975):

IF Attribute 1 is A_1 AND Attribute 2 is A_2 AND ... AND Attribute N is A_N, THEN Consequent is Ci,

In the expression, Attribute 1, Attribute 2, ... , Attribute N collectively form an Antecedent. Antecedents and Consequents are variables or concepts and $A1$, $A2$; ..., Ci are linguistic terms or fuzzy sets of these variables, such as, "low", "dangerous" or "high", etc. (Bandemer and Gottwald 1995).

For the situation of bushing diagnosis, the amalgamation of the combustible gases in the absence of oxygen does not generate a failure. With transformer oil, failure happens when oxygen is present in quantities above 0.2% at temperatures above 250°C without any spark being present (auto-ignition) or at 156°C if a spark is present (flash point). This condition decreases the number of fuzzy rules meaningfully, to only 81 fuzzy rules. The amalgamations were modeled in 81 compartments. Two examples of fuzzy rules in written language are (1) If hydrogen is High only then Low Risk and (2) If hydrogen is High and Oxygen is High then High Risk.

7.2.4 *Decisions Based on Fuzzy Rules*

By applying the rules, the bushing is given a risk rating for which certain maintenance actions must be taken on the plant. For the safe operation of bushings

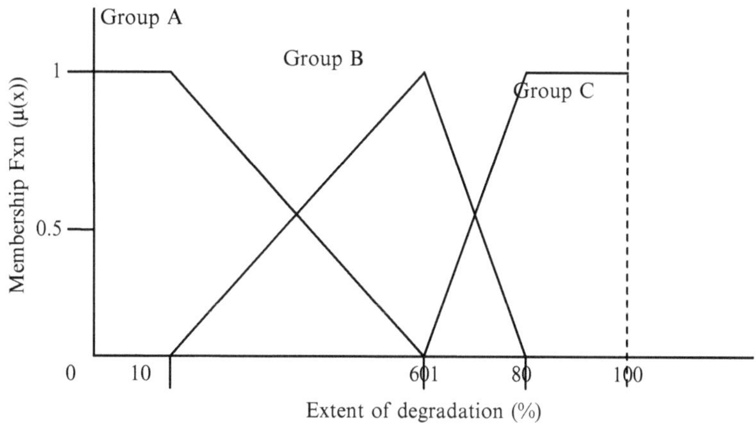

Fig. 7.3 Membership functions of decision

it is suggested that for all high risk situations, stop the transformer and remove the bushing from the transformer. For all medium risk situations, monitor the bushings more regularly. All low risk situations are allowed operate normally. From the decision table an aggregated membership is developed using the following equations (Bandemer and Gottwald 1995; Dhlamini et al. 2006; Dhlamini 2007):

$$\mu_{agg} = \mu_{LR} \cup \mu_{MR} \cup \mu_{HR} \tag{7.31}$$

Here

$$\mu_{LR}(x) = \begin{cases} 1 & \text{for } x \leq 10 \\ -0.01667x + 1 & \text{for } 10 \leq x \leq 60 \end{cases}$$

$$\mu_{MR}(x) = \begin{cases} 0.01667x - 1 & \text{for } 10 \leq x \leq 60 \\ -0.05x + 4 & \text{for } 60 \leq x \leq 80 \end{cases}$$

$$\mu_{HR}(x) = \begin{cases} 0.05x - 3 & \text{for } 60 \leq x \leq 80 \\ 1 & \text{for } x \geq 80 \end{cases} \tag{7.32}$$

The graph of the membership functions are shown in Fig. 7.3. The membership function is asymmetrical so that a decision to omit damaged bushings is stricter than that of slightly safe bushings. This means that, small alterations in a condition that is hazardous are underlined by the membership function. A steeper gradient on the graph permits the user to recognize those components which have small variances in critical levels of concentrations of hazardous gases.

7.2.5 Aggregated Rules

The table of fuzzy rules can additionally be streamlined by identifying cells with features that are common within partitions. This procedure is known as *aggregating*. One can improve the following Aggregated Rules (AR) based on the underlined partition (Dhlamini et al. 2006).

(AR4) IF bushing has 'Dangerous level of TDCG' AND 'NOT Normal Oxygen' AND 'Not Normal Methane', THEN the bushing belongs to 'Group A (high risk or dangerous)'.

(AR5) IF bushing has 'Dangerous TDCG' AND 'NOT Normal Oxygen' AND 'Normal Methane', THEN the bushing belongs to 'Group B (medium risk or elevated)'.

(AR6) IF bushing has 'Dangerous TDCG' AND 'Normal Oxygen' AND 'Not Normal Methane', THEN the bushing belongs to 'Group B (medium risk or elevated)'.

(AR7) IF bushing has 'Dangerous TDCG' AND 'Normal Oxygen' AND 'Normal Methane', THEN the bushing belongs to 'Group C (low risk or normal)'.

In rule AR1, the result is 'the bushing belongs to Group A'. The truth value of this consequence (CAR4) is displayed in Eq. 7.33 (Dhlamini 2007).

$$CAR_4 = \min(1, 1, 1) = 1 \tag{7.33}$$

When all the rules have been applied to a specific bushing, and different truth values of each consequence obtained, the *maximum* value of each consequence among all the rules that result in that consequence is taken as the degree to which that consequence applies to a given bushing. In the end this gives rise to an aggregated fuzzy output as shown in Table 7.1 and the following equation (Bandemer and Gottwald 1995):

$$AGD_i = \max(CAR1_i \cap CAR2_i \cap \cdots \cap CARn_i) \tag{7.34}$$

Here, AGD_i is the aggregated decision for category i, CAR is the consequence of aggregated rules in a particular category i in a given partition.

7.2.6 Defuzzification

Defuzzification is intended at transforming fuzzy information into crisp data. The technique used for defuzzification in this chapter is known as the *weighted average of maximum values of membership functions* (Majozi and Zhu 2005). The technique was chosen since it is computationally economical. The result of this technique offers the rank or level of risk of each bushing.

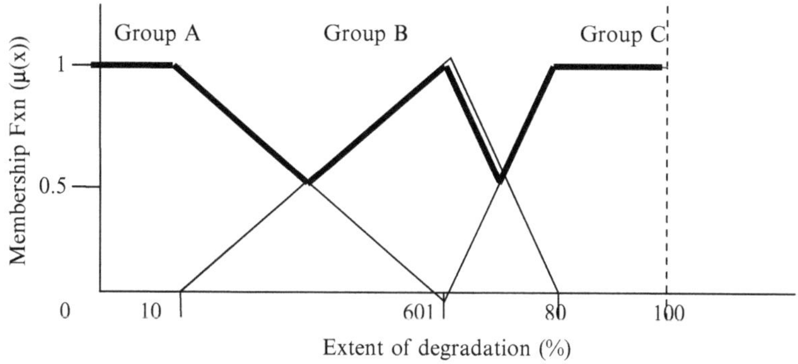

Fig. 7.4 Aggregated output

For one arbitrary bushing numbered 200323106, the rank as obtained using Eq. 7.35 and Fig. 7.4 shows the aggregated membership function from which the values in this equation were obtained (Majozi and Zhu 2005; Dhlamini et al. 2006; Dhlamini 2007).

$$Rank(200323106) = \frac{\left(\frac{0+10}{2}\right)\mu(Group_A(x)) + 60\,\mu(Group_B(x)) + \left(\frac{80+100}{2}\right)\mu(Group_C(x))}{\mu(Group_A(x)) + \mu(Group_B(x)) + \mu(Group_C(x))} \quad (7.35)$$

The coefficients shown in Eq. 7.35 are the levels of risk of failure corresponding to the maximum values of the respective sets.

7.3 Fuzzy ARTMAP

Fuzzy ARTMAP is a neural network structure based on Adaptive Resonance Theory (ART) that is capable of supervised learning of an arbitrary Mapping of clusters in the input space and their associated labels. The vital characteristic of this type of network structure is that it is capable of rapid, online, incremental learning, classification and prediction (Carpenter et al. 1992; Marwala 2009). Fuzzy ARTMAP has been successfully used by Lopes et al. (2005) for electricity load forecasting and Tan et al. (2008) used a fuzzy ARTMAP for conflict resolution. A fuzzy ARTMAP has been used in condition monitoring by Javadpour and Knapp (2003b) even though their application was not online. The Fuzzy ARTMAP structure applied in this chapter was comprised of two ART modules (ARTa and ARTb) that created stable recognition categories in response to sequences of the input pattern.

Nelwamondo and Marwala (2007) applied an ensemble of fuzzy ARTMAP for dealing with missing data without forecasting or estimating the missing values. Their method was found to be appropriate for online operations of neural networks and was used for online condition monitoring. Their method was tested both on classification and regression problems. An ensemble of fuzzy ARTMAPs was implemented for classification while an ensemble of multi-layer perceptions was applied for the regression problem. The results achieved using this ensemble-based technique were compared to those attained using a combination of auto-associative neural networks and genetic algorithms and they indicated that this technique performed up to 9% better in regression problems.

Vilakazi and Marwala (2006) applied a fuzzy ARTMAP method for intrusion detection and diagnosis. Their method applied a Sequential Backward Floating Search for feature selection and a fuzzy ARTMAP for detection and diagnosis of attacks. The optimal vigilance parameter for the fuzzy ARTMAP was chosen using a genetic algorithm. The reduced set of features decreased the computation time by 0.789 s. A classification rate of 100% and 99.89% was obtained for the detection stage and diagnosis stage, respectively.

Vilakazi and Marwala (2007a, b) applied a Fuzzy ARTMAP (FAM) for incremental learning for bushing condition monitoring. FAM was introduced since it can incrementally learn from information as it becomes available. An ensemble of classifiers was used to improve the classification accuracy of the systems. The results showed that the FAM ensemble gave an accuracy of 98.5%. Additionally, the results showed that the fuzzy ARTMAP can update its knowledge in an incremental fashion without forgetting previously learned information.

Barszcz et al. (2011) applied a fuzzy ART neural network for wind turbines state classification and Chang et al. (2010b) applied a fuzzy ART for color-based semantic image retrieval. Wang and Zan (2010) applied a fuzzy ART for statistical process control while Chen et al. (2010) successfully applied a fuzzy ART for infrared target detection. Other successful applications of fuzzy ART include its medical application in the diagnosis of cancer (Hwang et al. 2010), for semantic image retrieval (Chang et al. 2010a) and for personal credit scoring (Jiang and Lin 2010).

The architecture of the FAM is shown in Fig. 7.5. The two ART modules were interconnected by a series of weighted connections between the F2 layers. The connection between the two ART modules formed the Map Field (Carpenter et al. 1992). Various parameters needed to be set for the training process and this was achieved in the ART Network. The *vigilance parameter*, $\rho[0,1]$ is the only user-specified parameter. The vigilance parameter controls the network resonance. The second parameter that is adjusted during training is the *training rate* $\beta[0,1]$ which controls the adaptation speed. Here 0 implies a slow speed and 1 implies the fastest. The parameter α acts as a parameter to decide the category class and is always set such that $0 < \alpha < 1$.

During supervised learning, an analog signal vector $X = (a, a^c)$ was input into the ARTa input layer F_0^a in the form of a complement code. Both the original input a and its complement a^c were presented to the fuzzy ARTMAP network as explained by Javadpour and Knapp (2003b).

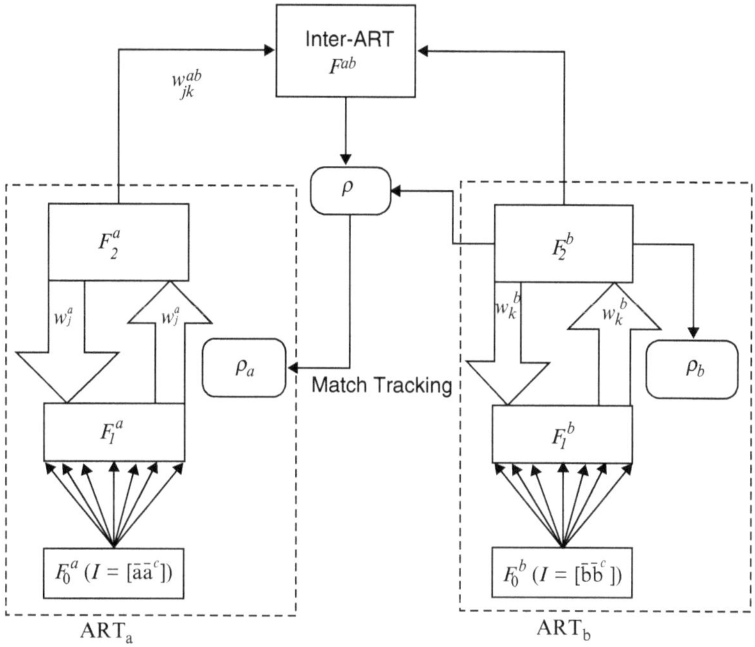

Fig. 7.5 Architecture of the fuzzy ARTMAP

Each component in the input vector corresponds to a single node in F_1^a. The key function of this F_1^a block was to compare the hypothesized class propagated from the F_2^a to the input signal. Simultaneously, a vector b was presented to *ARTb* and this vector contained the desired outputs corresponding to the vector a in the *ARTa*. The network then used hypothesis testing to deduce which category the input pattern should belong to. Mapping was done in two steps. Firstly, the *ARTa* module allowed data to be clustered into categories that were mapped to a class in the *ARTb* side of the module. The Fuzzy ARTMAP map field mapped the data cluster in the '*A*-side' to the label cluster in the '*B*-side'. During the learning process, each template from the '*A*-side' was mapped to one template on the '*B*-Side', ensuring a many-to-one mapping. The weights w_{jk}^{AB} were used to control the association between the *F2* nodes on both sides. When the vectors a and b were presented to *ARTa* and *ARTb* respectively, both models soon entered resonance.

When the vigilance criterion was respected, the map field learned the association between vectors a and b by modifying its weights following the initial weight. A fuzzy ARTMAP has an internal controller that ensures autonomous system operation in real time. The inter-ART module has a self-regulatory mechanism named *match tracking*, whose objective was to maximize the generalization and minimize the network error. A complete description of the Fuzzy ARTMAP is

Table 7.1 Classification results

Classifier	Validation accuracy (%)	Test accuracy (%)
Fuzzy set network	98.4	98.0
Fuzzy ARTMAP	98.5	97.5

provided by Carpenter et al. (1992). The vigilance criterion is given by (Carpenter et al. 1992; Marwala 2009):

$$\frac{\left|y^b \Lambda w_{JK}^{ab}\right|}{y_b} \geq \rho_{ab} \tag{7.36}$$

Here the H vigilance criterion was reached by making a small increment in the vigilance parameter such that a certain category was excluded. This was done until the moment the active category corresponded to the desired output. After the input had completed the resonance state by the vigilance criterion, the weight adaptation was accomplished. The adaptation of the *ARTa* and *ARTb* module weights is given by (Carpenter et al. 1992; Marwala 2009):

$$w_j^{new} \beta(I \Lambda w_j^{old}) + (1-\beta) w_j^{old} \tag{7.37}$$

7.4 Results

Fuzzy sets technique and fuzzy ARTMAP were implemented to classify faults in transformer bushings using the dissolved gas analysis data based on the IEC60599, IEEE C57-104, and the California State University Sacramento (CSUS) criteria for Oil Impregnated Paper (OIP) bushings, described in Chapter 2 by Vilakazi (2007). The networks were trained, validated and tested with 1,000 data points that consisted of 500 faulty and 500 healthy bushings. The results are shown in Table 7.1.

Table 7.1 shows that the fuzzy set network gave marginally better results than the fuzzy ARTMAP. However, the fuzzy set network was, however, able to factor into account the user experience through the fuzzy rules.

7.5 Conclusion

In this chapter fuzzy set theory and fuzzy ARTMAP were implemented for fault identification in transformer bushings based on the dissolved gas analysis (DGA) data and the IEC60599, IEEE C57-104, and California State University Sacramento (CSUS) criteria. The results showed that the fuzzy set theory and the fuzzy ARTMAP gave an accuracy of 98% and 97.5%, respectively. The fuzzy set theory can incorporate user experience through the use of fuzzy rules.

References

Aliustaoglu C, Metin Ertunc H, Ocak H (2009) Tool wear condition monitoring using a sensor fusion model based on fuzzy inference system. Mech Syst Signal Process 23(2):539–546

Ammar S, Wright R (2000) Applying fuzzy-set theory to performance evaluation. Socioecon Plann Sci 34:285–302

Araujo E (2008) Improved Takagi-Sugeno fuzzy approach. In: Proceedings of the IEEE international conference on fuzzy systems, Hong Kong, China, pp 1154–1158

Babuska R (1991) Fuzzy modeling and identification. PhD thesis, Technical University of Delft

Bandemer H, Gottwald S (1995) Fuzzy sets, fuzzy logic, fuzzy methods with applications. Wiley, New York

Barszcz T, Bielecka M, Bielecki A, Wójcik M (2011) Wind turbines states classification by a fuzzy-ART neural network with a stereographic projection as a signal normalization. Lect Notes Comput Sci (including subseries Lecture Notes in Artificial Intelligence and Lecture Notes in Bioinformatics) 6594 LNCS (PART 2):225–234

Biacino L, Gerla G (2002) Fuzzy logic, continuity and effectiveness. Arch Math Logic 41:643–667

Bih J (2006) Paradigm shift – an introduction to fuzzy logic. IEEE Potentials 25:6–21

Boesack CD, Marwala T, Nelwamondo FV (2010) Application of GA-fuzzy controller design to automatic generation control. In: 2010 Third international workshop on advanced computational intelligence (IWACI), Suzhou, China, pp 227–232

Bojadziev G, Bojadziev M (1995) Fuzzy sets, fuzzy logic, applications. World Scientific Publishing Co. Pte. Ltd, Singapore/River Edge

Cabalar AF, Cevik A, Gokceoglu C, Baykal G (2010) Neuro-fuzzy based constitutive modeling of undrained response of Leighton Buzzard sand mixtures. Expert Syst Appl 37:842–851

Cantor G (1874) Über eine Eigenschaft des Inbegriffes aller reellen algebraischen Zahlen. Crelles J Math 77:258–262

Carpenter GA, Grossberg S, Markuzon N, Reynolds JH, Rosen DB (1992) Fuzzy ARTMAP: a neural network architecture for incremental supervised learning of analog multidimensional maps. IEEE Trans Neural Netw 3:698–713

Catelani M, Fort A, Alippi C (2002) A fuzzy approach for soft fault detection in analog circuits. Measurement 32(1):73–83

Chang CY, Wang HJ, Jian RH (2010a) Semantic image retrieval with Fuzzy-ART. In: 2010 International conference on system science and engineering, ICSSE 2010, Taipei, Taiwan, art.# 5551705, pp 69–74

Chang CY, Wang HJ, Jian RH (2010b) Color-based semantic image retrieval with fuzzy-ART. In: Proceedings of 2010 6th international conference on intelligent information hiding and multimedia signal processing, IIHMSP 2010, Darmstadt, Germany, art. #5635903, pp 426–429

Chen B, Wang W, Qin Q (2010) Infrared target detection based on fuzzy ART neural network. In: 2010 2nd international conference on computational intelligence and natural computing, CINC 2010, 2, Wuhan, China, art. no. 5643745, pp 240–243

Chen J, Roberts C, Weston P (2008) Fault detection and diagnosis for railway track circuits using neuro-fuzzy systems. Control Eng Pract 16(5):585–596

Cox E (1994) The fuzzy systems handbook: a practitioner's guide to building, using, maintaining fuzzy systems. AP Professional, Boston

D'Angelo MFSV, Palhares RM, Takahashi RHC, Loschi RH, Baccarini LMR, Caminhas WM (2011) Incipient fault detection in induction machine stator-winding using a fuzzy-Bayesian change point detection approach. Appl Software Comput 11(1):179–192

Demirli K, Khoshnejad M (2009) Autonomous parallel parking of a car-like mobile robot by a neuro-fuzzy sensor-based controller. Fuzzy Sets Syst 160:2876–2891

Devlin K (1993) The joy of sets. Springer, Berlin

Dhlamini SM (2007) Bushing diagnosis using artificial intelligence and dissolved gas analysis. University of the Witwatersrand PhD thesis

Dhlamini SM, Marwala T (2005) Modeling inaccuracies from simulators for HV polymer Bushing. In: Proceedings of international symposium on high voltage, Beijing, Paper A18
Dhlamini SM, Marwala T, Majozi T (2006) Fuzzy and multilayer perceptron for evaluation of HV bushings. In: Proceedings of the IEEE international conference on systems, man and cybernetics, Taiwan, pp 1331–1336
El-Sebakhy EA (2010) Flow regimes identification and liquid-holdup prediction in horizontal multiphase flow based on neuro-fuzzy inference systems. Math Comput Simul 80:1854–1866
Evsukoff A, Gentil S (2005) Recurrent neuro-fuzzy system for fault detection and isolation in nuclear reactors. Adv Eng Inform 19(1):55–66
Ferreirós J (1999) Labyrinth of thought: a history of set theory and its role in modern mathematics. Birkhäuser, Basel
Flaig A, Barner KE, Arce GR (2000) Fuzzy ranking: theory and applications. Signal Process 80:1017–1036
Hájek P (1995) Fuzzy logic and arithmetical hierarchy. Fuzzy Sets Syst 3:359–363
Hájek P (1998) Metamathematics of fuzzy logic. Kluwer, Dordrecht
Halpern JY (2003) Reasoning about uncertainty. MIT Press, Cambridge
Hsu Y-C, Lin S-F (2009) Reinforcement group cooperation-based symbiotic evolution for recurrent wavelet-based neuro-fuzzy systems. J Neurocomput 72:2418–2432
Huang T-M, Kecman V (2005) Gene extraction for cancer diagnosis by support vector machinesan improvement. Artif Intell Med 35:185–194
Hwang JIG, Liu CE, Sokoll L, Adam BL (2010) Applying fuzzy ART in medical diagnosis of cancers. In: 2010 international conference on machine learning and cybernetics, ICMLC 2010, Qingdao, 3, art. no. 5580939, pp 1084–1089
Iplikci S (2010) Support vector machines based neuro-fuzzy control of nonlinear systems. J Neurocomput 73:2097–2107
Javadpour R, Knapp GM (2003a) A fuzzy neural network approach to condition monitoring. Comput Ind Eng 45:323–330
Javadpour R, Knapp GM (2003b) A fuzzy neural network approach to machine condition monitoring. Comput Ind Eng 45(2):323–330, 25th international conference on computers and industrial engineering, Limerick, Ireland, August 2003
Jeffries M, Lai E, Plantenberg DH, Hull JB (2001) A fuzzy approach to the condition monitoring of a packaging plant. J Mater Process Technol 109(1–2):83–89
Jiang M, Lin S (2010) A study of personal credit scoring models based on fuzzy ART. J Comput Info Syst 6(9):2805–2811
Johnson P (1972) A history of set theory. Prindle, Weber & Schmidt, Boston
Klir GJ, Folger TA (1988) Fuzzy sets, uncertainty, and information. Prentice Hall, Englewood Cliffs
Klir GJ, Yuan B (1995) Fuzzy sets and fuzzy logic: theory and applications. Prentice Hall, Upper Saddle River
Klir GJ, St Clair UH, Yuan B (1997) Fuzzy set theory: foundations and applications. Prentice Hall, Upper Saddle River
Korbicz J, Kowal M (2007) Neuro-fuzzy networks and their application to fault detection of dynamical systems. Eng Appl Artif Intell 20(5):609–617, Soft Computing Applications, August 2007
Kosko B (1993) Fuzzy thinking: the new science of fuzzy logic. Hyperion, New York
Kosko B, Isaka S (1993) Fuzzy logic. Sci Am 269:76–81
Kubica EG, Wang D, Winter AD (1995) Modelling balance and posture control mechanisms of the upper body using conventional and fuzzy techniques. Gait Posture 3(2):111
Lau HCW, Dwight RA (2011) A fuzzy-based decision support model for engineering asset condition monitoring – a case study of examination of water pipelines. Expert Syst Appl 38(10):13342–13350
Lopes MLM, Minussi CR, Lotufo ADP (2005) Electric load forecasting using a fuzzy ART&ARTMAP neural network. Appl Software Comput 5(2):235–244

Majozi T, Zhu XX (2005) A combined fuzzy set theory and MILP approach in integration of planning and scheduling of batch plants – personnel evaluation and allocation. Comput Chem Eng 29:2029–2047, Elsevier Science, July 2005

Mamdani EH (1974) Application of fuzzy algorithms for the control of a dynamic plant. Proc IEE 121:1585–1588

Mamdani EH, Assilian S (1975) An experiment in linguistic synthesis with a fuzzy logic controller. Int J Man–Mach Stud 7(1):1–13

Marwala T (2009) Computational intelligence for missing data imputation, estimation and management: knowledge optimization techniques. IGI Global Publications, Information Science Reference Imprint, New York

Marwala T, Lagazio M (2011) Militarized conflict modeling using computational intelligence techniques. Springer, London

Mendonca LF, Sousa JMC, Sá da Costa JMG (2009) An architecture for fault detection and isolation based on fuzzy methods. Expert Syst Appl 36(2, Part 1):1092–1104

Msiza IS, Szewczyk M, Halinka A, Pretorius J-HC, Sowa P, Marwala T (2011) Neural networks on transformer fault detection: evaluating the relevance of the input space parameters. In: 2011 IEEE/PES Power Systems Conference and Exposition, PSCE 2011, Phoenix, art. no. 5772567

Nelwamondo FV, Marwala T (2007) Fuzzy artmap and neural network approach to online processing of inputs with missing values. Trans S Afr Inst Electrical Eng 98(2):45–51

Novák V (1989) Fuzzy sets and their applications. Adam Hilger, Bristol

Novák V (2005) On fuzzy type theory. Fuzzy Sets Syst 149:235–273

Novák V, Perfilieva I, Močkoř J (1999) Mathematical principles of fuzzy logic. Kluwer, Dordrecht

Razavi-Far R, Davilu H, Palade V, Lucas C (2009) Model-based fault detection and isolation of a steam generator using neuro-fuzzy networks. Neurocomputing 72(13–15):2939–2951, Hybrid Learning Machines (HAIS 2007)/Recent Developments in Natural Computation (ICNC 2007), August 2009

Sainz Palmero GI, Juez Santamaria J, Moya de la Torre EJ, Peran Gonzalez JR (2005) Fault detection and fuzzy rule extraction in AC motors by a neuro-fuzzy ART-based system. Eng Appl Artif Intell 18(7):867–874

Sugeno M (1985) Industrial applications of fuzzy control. Elsevier, Amsterdam

Sugeno M, Kang G (1988) Structure identification of fuzzy model. Fuzzy Sets Syst 28:15–33

Takagi T, Sugeno M (1985) Fuzzy identification of systems and its applications to modeling and control. IEEE Trans Syst Man Cybern 15:116–132

Tan SC, Rao MVC, Lim Fuzzy CP (2008) ARTMAP dynamic decay adjustment: an improved fuzzy ARTMAP model with a conflict resolving facility. Appl Software Comput 8(1):543–554

Tettey T, Marwala T (2006) Neuro-fuzzy modeling and fuzzy rule extraction applied to conflict management. Lect Notes Comput Sci 4234:1087–1094

Vilakazi CB (2007) Machine condition monitoring using artificial intelligence: the incremental learning and multi-agent system approach, University of the Witwatersrand Masters dissertation

Vilakazi CB, Marwala T (2006) Application of feature selection and fuzzy ARTMAP to intrusion detection. In: IEEE international conference on systems, man and cybernetics, Vancouver, Canada, pp 4880–4885

Vilakazi CB, Marwala T (2007a) Incremental learning and its application to bushing condition monitoring. Lect Notes Comput Sci (including subseries Lecture Notes in Artificial Intelligence and Lecture Notes in Bioinformatics) 4491 LNCS (PART 1):1237–1246

Vilakazi CB, Marwala T (2007b) Online incremental learning for high voltage bushing condition monitoring. In: IEEE international conference on neural networks – conference proceedings, Orlando, Florida, art. no. 4371355, pp 2521–2526

Von Altrock C (1995) Fuzzy logic and neurofuzzy applications explained. Prentice Hall, Englewood Cliffs

Wang M, Zan T (2010) Adaptively pattern recognition in statistical process control using fuzzy ART neural network. In: Proceedings – 2010 international conference on digital manufacturing and automation, ICDMA 2010, 1, Changcha, China, art. no. 5701122, pp 160–163

Wang Z (2000) Artificial intelligence applications in the diagnosis of power transformer incipient faults, PhD thesis, Virginia Polytechnic Institute and State University
Wright S, Marwala T (2006) Artificial intelligence techniques for steam generator modelling. arXiv:0811.1711
Zadeh LA (1965) Fuzzy sets. Info Control 8:338–353
Zadeh LA (1973) Outline of a new approach to the analysis of complex systems and decision processes. IEEE Trans Syst Man Cybern 3(1):28–44
Zemankova-Leech M (1983) Fuzzy relational data bases. PhD dissertation, Florida State University
Zimmermann H (2001) Fuzzy set theory and its applications. Kluwer, Boston

Chapter 8
Rough Sets for Condition Monitoring

8.1 Introduction

Rough set theory as put forward by Pawlak (1991) is a mathematical technique which models vagueness and uncertainty. It permits one to estimate sets that are difficult to explain even with accessible information. For this chapter, rough sets were applied to the condition monitoring of transformer bushings based on dissolved gas analysis data. The advantages of rough sets, as with many other computational intelligence methods, are that they do not necessitate inflexible *a priori* assumptions about the mathematical characteristics of such complex relationships, as normally required for the multivariate statistical approaches (Machowski and Marwala 2005; Crossingham et al. 2008; Marwala and Lagazio 2011). Rough set theory is based on the assumption that the information of interest is associated with some information from its universe of discourse (Crossingham and Marwala 2008a, b; Tettey et al. 2007; Marwala and Crossingham 2008, 2009; Crossingham et al. 2009).

Wang et al. (2006) applied rough set theory to handle uncertainty and thereby decreased the redundancy in evaluating the degree of malignancy in brain glioma, based on Magnetic Resonance Imaging findings as well as the clinical data before an operation. Their data comprised unsuitable features, uncertainties and missing values. The rough set rules that were identified from these data were applied to evaluate the degree of the malignancy. Rough set based feature selection procedures were used to select features so that the accuracy of classification based on decision rules could be enhanced. These selected feature subsets were applied to give decision rules for the classification task. The results obtained showed that their technique identified reducts that produced decision rules with higher classification rates than conventional methods.

Xie et al. (2011) used variable precision rough set for land use / land cover retrieval from remote sensing images. Their results showed a retrieval accuracy of 87.32%. Chen et al. (2011) used a rough set method for the prediction of protein interaction hot spots. Their results indicated that four features, viz. the change of

T. Marwala, *Condition Monitoring Using Computational Intelligence Methods: Applications in Mechanical and Electrical Systems*, DOI 10.1007/978-1-4471-2380-4_8,

accessible surface area, percentage of the change of accessible surface area, size of a residue, and atomic contacts were vital in predicting the hot spots.

Salamó and López-Sánchez (2011) used rough sets for selecting features in Case-Based Reasoning classifiers. Lin et al. (2011) applied a hybrid of rough set theory and flow network graphs to predict the customer churn in credit card accounts using 21,000 customer samples equally divided into three classes (survival, voluntary churn, and involuntary churn). The input data included demographic, psychographic, and transactional variables for studying and classifying customer characteristics. Their results indicated that rough sets can forecast customer churn and offer valuable information for decision-makers.

Other applications of rough set theory include the work by Azadeh et al. (2011) who applied a rough set technique for assessing the efficiency of personnel, Zhang et al. (2010) for controlling reagents in an ionic reverse flotation process, Huang et al. (2011) who used rough sets in patent development with the emphasis on resource allocation, Zou et al. (2011) who used rough sets for distributor selection in a supply chain management system, Wang et al. (2010) who used rough sets and a Tabu search for credit scoring, Gong et al. (2010) for a rare-earth extraction process, Chen et al. (2010a) for creating a diagnostic system based on Chinese traditional medicine for the pneumonia in elderly, Yan et al. (2010) for predicting soil moisture, and Liao et al. (2010) for a model that assessed brand trust. The main concept of rough set theory is an indiscernibility relation, where indiscernibility indicates indistinguishable from one another. For knowledge attainment from data with numerical attributes, special methods are applied. Most commonly a step called *discretization* is taken before the main step of rule induction or decision tree generation is applied (Crossingham and Marwala 2007). A number of approaches to achieve the goal of discretization are Boolean reasoning, Equal-Width-Bin (EWB) partitioning and Equal-Frequency-Bin (EFB) partitioning (Jaafar et al. 2006; Fayyad and Irani 1993).

For this chapter an ant colony optimization method and an EFB partitioning method was used to discretize the rough set partitions and apply these to condition the monitoring of transformer bushings (Mpanza and Marwala 2011). Zhao et al. (2010) applied a fuzzy preference based rough set technique and principal component analysis for condition monitoring. They implemented a Principal Component Analysis (PCA) to reduce the input space. Their method was tested for damage level detection of an impeller in a slurry pump. Zhang et al. (2008) applied rough sets and mathematical morphology for intelligent condition monitoring. They used the theory of image processing to analyze the flank faces and they used tool condition monitoring through measuring the area of tool wear. Their results indicated that their method was flexible and fast enough to be applied in real time for the condition monitoring of tool wear online. Li et al. (2005) applied rough sets for the condition monitoring of an engine. They created decision tables for each fault case and used a rough set reduction to conduct intelligent fault diagnosis and obtained good results.

Shen et al. (2000) applied rough set theory for the fault diagnosis of a multi-cylinder diesel engine. When the reducts from the rough sets theory were analyzed it was observed that this method was effective for valve fault diagnosis. In addition a new discretization technique was presented and was found to be suitable for discretizing the attributes without *a priori* knowledge.

Xiang et al. (2009) applied the Walsh transform and rough sets for fault diagnosis. Data processed by the Walsh transform was discretized and reduced by the rough sets theory, and diagnosis rules were extracted and used for fault diagnosis. The results obtained showed a higher accuracy than other methods.

Wang and Li (2004) successfully applied a rough-set based fault-ranking prototype system for fault diagnosis and achieved good results. Elsewhere, Tay and Shen (2003) applied rough set for fault diagnosis in a multi-cylinder diesel engine.

Other applications of rough set theory to the area of condition monitoring include that for a mono-block centrifugal pump (Sakthivel et al. 2010), for fault line detection of ineffectually grounded systems (An et al. 2011), nuclear power plants (Mu et al. 2011), in fault line detection for distribution networks (Pang et al. 2010), machines (Yu and Han 2010), and in diesel engines (Li et al. 2010).

8.2 Rough Sets

The primary objective of using rough sets is to produce estimates of various concepts from the acquired data. Contrary to other approaches that are applied to handle uncertainty, rough set theory has its own exclusive benefits in that it does not necessitate (Crossingham 2007; Nelwamondo 2008; Marwala and Lagazio 2011):

- any additional information about the experimental training data such as the statistical probability; and
- basic probability assignment in fuzzy set theory (Pawlak and Munakata 1996).

Rough set theory deals with the estimation of sets that are hard to explain with the available information (Ohrn 1999; Ohrn and Rowland 2000; Marwala and Lagazio 2011). It is targeted mainly to the classification of imprecise, uncertain, or incomplete information. Two estimations, the upper and lower estimation are developed to handle inconsistent information. The data are represented using an information table.

Rough set theory is based on a set of rules, which are described in terms of linguistic variables. Rough sets are of essential significance to artificial intelligence and cognitive science, and are well applied to the tasks of machine learning and decision analysis, particularly in the analysis of decisions in which there are contradictions. Because they are rule-based, rough sets are highly transparent but they are not as accurate. However, they are not good as universal estimators, since other machine learning techniques such as neural networks are better in their predictions. Thus, in machine learning, there is always a trade-off between prediction accuracy and transparency.

Crossingham and Marwala (2007) presented a method to optimize the partition sizes of rough set using various optimization methods. Three optimization methods were applied to perform the granularization process: the genetic algorithm, hill climbing, and simulated annealing. These optimization approaches maximize the classification accuracy of the rough sets. Their rough set partition approaches were tested on a demographic set. The three methods were compared for their computational time, accuracy, and number of rules produced and then applied to a HIV data set. The optimized technique results were then compared to a non-optimized discretization method, using Equal-Width-Bin (EWB) partitioning. The accuracies achieved after optimizing the partitions using a genetic algorithm (GA), hill climbing, and simulated annealing (SA) were 66.89%, 65.84%, and 65.48%, respectively, compared to the accuracy of the EWB partitioning of 59.86%. In addition to rough sets providing the plausibility of the estimated HIV status, they also provided the linguistic rules describing how demographic parameters drive the risk of HIV.

Rough set theory offers a method of reasoning from vague and imprecise data (Goh and Law 2003). The method is based on the assumption that some observed information is in some way associated with some information in the universe of the discourse (Komorowski et al. 1999; Yang and John 2006; Kondo 2006). This suggests that if some characteristics of the data are missing, then they can be approximated from part of the information in the universe of discourse which is comparable with the observed part of that specific data. Objects with the same information are indiscernible in the view of the available information. An elementary set consisting of indiscernible objects forms a basic granule of knowledge. A union of an elementary set is referred to as a *crisp set*; or else, the set is considered to be *rough*. In the next sub-sections, rough set theory is described.

8.2.1 Information System

An information system (Λ), is described as a pair (U, A) where U is a finite set of objects known as the universe and A is a non-empty finite set of attributes as described as follows (Crossingham 2007; Yang and John 2006; Nelwamondo 2008; Marwala 2009; Marwala and Lagazio 2011).

$$\Lambda = (U, A) \tag{8.1}$$

All attributes $a \in A$ have values, which are elements of a set V_a of the attributes a (Dubois 1990; Crossingham 2007; Marwala and Lagazio 2011):

$$a : U \to V_a \tag{8.2}$$

A rough set is described with a set of attributes and the indiscernibility relation between them. Indiscernibility is explained in the next subsection.

8.2.2 The Indiscernibility Relation

The *indiscernibility relation* is one of the central ideas of rough set theory (Grzymala-Busse and Siddhaye 2004; Zhao et al. 2007; Pawlak and Skowron 2007; Marwala and Lagazio 2011). *Indiscernibility* basically suggests similarity (Goh and Law 2003) and, consequently, these sets of objects are indistinguishable. Given an information system Λ and subset$B \subseteq A$, B the indiscernibility defines a binary relation *I(B)* on U such that (Pawlak et al. 1988; Ohrn 1999; Wu et al. 2003; Ohrn and Rowland 2000; Nelwamondo 2008; Marwala and Lagazio 2011):

$$\begin{aligned} &(x, y) \in I(B) \\ &\textit{if and only if} \\ &a(x) = a(y) \end{aligned} \tag{8.3}$$

for all $a \in A$ where $a(x)$ symbolizes the value of attribute a for element x. Equation 8.3 suggests that any two elements that are elements of $I(B)$ should be identical from the point of view of a. Supposing that U has a finite set of N objects $\{x_1, x_2, .., x_N\}$. Let Q be a finite set of n attributes $\{q_1, q_2, .., q_n\}$ in the same information system Λ, then (Inuiguchi and Miyajima; 2007; Crossingham 2007; Nelwamondo 2008; Marwala and Lagazio 2011):

$$\Lambda = \langle U, Q, V, f \rangle \tag{8.4}$$

Here f is the total decision function, known as the *information function*. From the explanation of the indiscernibility relation, two entities have a similarity relation to attribute a if they universally have the same attribute values.

8.2.3 Information Table and Data Representation

An information table is applied in rough sets theory as a technique for signifying the data. Data in the information table are organized, centered on their condition attributes and decision attributes (D). *Condition attributes* and *decision attributes* are similar to the independent variables and dependent variable (Goh and Law 2003). These attributes are divided into $C \cup D = Q$ and $C \cup D = 0$. Data is indicated in the table and each object is characterized in an *Information System* (Komorowski et al. 1999).

8.2.4 Decision Rules Induction

Rough sets also require producing decision rules for a given information table. The rules are generally based on condition attributes values (Bi et al. 2003; Slezak and

Ziarko 2005). The rules are presented in an '*if CONDITION(S)-then DECISION*' format. Stefanowski (1998) applied a rough set technique for inference in decision rules.

8.2.5 The Lower and Upper Approximation of Sets

The lower and upper approximations of sets are defined on the basis of the indiscernibility relation. The *lower approximation* is defined as the collection of cases whose equivalent classes are confined in the cases that need to be estimated, while the *upper approximation* is defined as the collection of classes that are incompletely contained in the set that need to be estimated (Rowland et al. 1998; Degang et al. 2006; Witlox and Tindemans 2004). If the concept X is defined as a set of all cases defined by a specific value of the decision and that any finite union of elementary set, associated with B called a B-definable set (Grzymala-Busse and Siddhaye 2004) then set X can be estimated by two B-definable sets, known as the B-lower estimation denoted by X and B-upper approximation $\overline{B}X$. The B-lower approximation is defined as (Bazan et al. 2004; Crossingham 2007; Nelwamondo 2008; Marwala and Lagazio 2011):

$$\underline{B}X = \{x \in U \,|[x]_B \subseteq X\} \tag{8.5}$$

and the B-upper approximation is defined as (Crossingham 2007; Nelwamondo 2008; Marwala and Lagazio 2011):

$$\overline{\mathrm{B}}X = \{x \in U \,|[x]_B \cap X \neq 0\} \tag{8.6}$$

There are other approaches that have been described for defining the lower and upper approximations for a completely specified decision table. Some of the popular ones include approximating the lower and upper approximation of X using Eqs. 8.7 and 8.8, as follows (Grzymala-Busse 2004; Crossingham 2007; Nelwamondo 2008; Marwala and Lagazio 2011):

$$\cup\{[x]_B \,|x \in U, [x]_B \subseteq X\} \tag{8.7}$$

$$\cup\{[x]_B \,|x \in U, [x]_B \cap X \neq 0\} \tag{8.8}$$

The definition of definability is revised in situations of incompletely specified tables. In this case, any finite union of characteristic sets of B is called a *B-definable set*. Three different definitions of approximations have been discussed by Grzymala-Busse and Siddhaye (2004). By letting B be a subset of A of all attributes and *R(B)* be the characteristic relation of the incomplete decision table with characteristic sets *K(x)*, where $x \in U$, the following can be defined (Grzymala-Busse 2004; Crossingham 2007; Nelwamondo 2008; Marwala and Lagazio 2011):

$$\underline{B}X = \{x \in U \,|K_B(x) \subseteq X\} \tag{8.9}$$

and

$$\overline{B}X = \{x \in U \,|K_B(x) \cap X \neq 0\} \tag{8.10}$$

Equations 8.9 and 8.10 are known as *singletons*. The subset lower and upper approximations of incompletely specified data sets can then be mathematically defined as (Nelwamondo 2008; Marwala and Lagazio 2011):

$$\cup\{K_B(x)\,|x \in U, K_B(x) \subseteq X\} \tag{8.11}$$

and

$$\cup\{K_B(x)\,|x \in U, K_B(x) \cap X = 0\} \tag{8.12}$$

Additional information on these approaches can be found in (Grzymala-Busse and Hu 2001; Grzymala-Busse and Siddhaye 2004; Crossingham 2007). It can be deduced from these properties that a crisp set is only defined if $\underline{B}(X) = \overline{B}(X)$. *Roughness* is consequently defined as the difference between the upper and the lower approximation.

8.2.6 Set Approximation

A number of properties of rough sets have been presented in the work of Pawlak (1991). An important property of rough set theory is the definability of a rough set (Quafafou 2000). This was explained for the situation when the lower and upper approximations are equal. If this is not the situation, then the target set is un-definable. Some of the distinctive cases of definability are (Pawlak et al. 1988; Crossingham 2007; Nelwamondo 2008; Marwala 2009):

- *Internally definable* set: Here, $\underline{B}X \neq 0$and $\overline{B}X = U$. The attribute set *B* has objects that certainly are elements of the target set *X*, even though there are no objects that can definitively be excluded from the set *X*.
- *Externally definable* set: Here, $\underline{B}X = 0$ and $\overline{B}X \neq U$. The attribute set *B* has no objects that certainly are elements of the target set *X*, even though there are objects that can definitively be excluded from the set *X*.
- *Totally un-definable* set: Here, $\underline{B}X = 0$and $\overline{B}X = U$. The attribute set *B* has no objects that certainly are elements of the target set *X*, even though there are no objects that can definitively be excluded from the set *X*.

8.2.7 The Reduct

An additional property of rough sets is the *reduct* which is a concept that defines whether there are attributes B in the information system that are more significant to the knowledge represented in the equivalence class structure than other attributes. It is vital to identify whether there is a subset of attributes which could be completely described by the knowledge in the database. This attribute set is known as the *reduct.*

Beynon (2001) concluded that the elementary feature of the variable precision rough set model involved an exploration for subsets of condition attributes which give identical information for classification functions as the complete set of given attributes. Beynon characterized these subsets as *approximate reducts.* Beynon explained these subsets for an identified classification error represented by β and then identified the particular variances and showed interesting consequences for identifying β-reducts which ensure a general knowledge similar to that obtained from the full set of attributes.

Terlecki and Walczak (2007) described the relations between rough set reducts and emerging patterns. Their study established a practical application for these observations for the minimal reduct problem, using these to test the differentiating factor of an attribute set. Shan and Ziarko (1995) properly defined a *reduct* as a subset of attributes $RED \subseteq B$ such that:

- $[x]_{RED} = [x]_B$. That is, the equivalence classes that were induced by reducing the attribute set *RED* are equal to the similar class structure that was induced by the full attribute set *B*.
- Attribute set *RED* is minimal because $[x]_{(RED-A)} \neq [x]_B$ for any attribute $A \in RED$. Simply, there is no attribute that can be taken away from the set *RED* without changing the equivalent classes $[x]_B$.

Therefore a reduct can be visualized as a suitable set of features that can adequately express the category's structure. One property of a reduct in an information system is that it is not unique since there may be other subsets of attributes which may still preserve the equivalence class structure conveyed in the information system. The set of characteristics that are common in all reducts is called a *core.*

8.2.8 Boundary Region

The *boundary region*, which can be expressed as the difference $\overline{B}X - \underline{B}X$, is a region which is composed of objects that cannot be included nor excluded as elements of the target set *X*. Simply, the lower approximation of a target set is an estimation which consists only of those objects which can be positively identified as elements of the set. The upper approximation is a rough approximation and includes objects that may be elements of the target set. The boundary region is the area between the upper and lower approximation.

8.2.9 Rough Membership Functions

A *rough membership function* is a function $\mu_A^x : U \rightarrow [0, 1]$ that, when applied to object x, quantifies the degree of overlap between set X and the indiscernibility set to which x belongs. The rough membership function is applied to estimate the plausibility and can be defined as (Pawlak 1991; Crossingham 2007; Nelwamondo 2008; Marwala and Lagazio 2011):

$$\mu_A^x(X) = \frac{|[x]_B \cap X|}{|[x]_B|} \tag{8.13}$$

The rough membership function can be understood as a fuzzification within the context of rough approximation. It *confirms* the translation from rough approximation into membership function. The important aspect of a rough membership function is that it *is* derived from data (Hoa and Son 2008; Crossingham 2007).

8.3 Discretization Methods

The methods which allow continuous data to be processed involve discretization. There are several methods available to perform discretization, but the two popularly ones – Equal-Width-Bin (EWB) partitioning and Equal-Frequency-Bin (EFB) partitioning – were investigated by Crossingham (2007). Details are given below.

8.3.1 Equal-Width-Bin (EWB) Partitioning

EWB partitioning divides the range of observed values of an attribute into$k$equally sized bins (Crossingham et al. 2009). For this chapter, $k$was taken as four. One notable problem of this method is that it is vulnerable to outliers that may drastically skew the data range. This problem was eliminated through a pre-processing step involving cleaning of the data. The manner in which data can be discretized using EWB follows (Grzymala-Busse 2004; Crossingham et al. 2009; Marwala and Lagazio 2011):

- Evaluate the Smallest and Largest value for each attribute and label these values S and L.
- Write the width of each interval, W, as:

$$W = \frac{L - S}{4} \tag{8.14}$$

- The interval boundaries can be determined as: $S + W$, $S + 2W$, $S + 3W$. These boundaries can be determined for any number of intervals k, up to the term $S + (k - 1)W$.

8.3.2 Equal-Frequency-Bin (EFB) Partitioning

EFB partitioning sorts the values of each attribute in ascending order and divides them into$k$bins where (given$m$instances) each bin contains *m*/*k* adjacent values. In most instances duplicated values will probably exist. The EFB partitioning can be implemented as follows (Grzymala-Busse 2004; Crossingham 2007; Marwala and Lagazio 2011):

- Arrange the values of each attribute $(v_1^a, v_2^a, v_3^a, ..., v_m^a)$ into intervals whereby m is the number of instances.
- Therefore each interval is made of the following sequential values:

$$\lambda = \frac{m}{4} \tag{8.15}$$

- The cut-off points may be computed using the following equation which is valid for $i = 1, 2, 3$ where k intervals can be calculated for $i = 1, \ldots, k$-1:

$$c_i = \frac{v_{i\lambda} + v_{i\lambda+1}}{2} \tag{8.16}$$

8.4 Rough Set Formulation

The process of modeling the rough set can be classified into these five stages (Grzymala-Busse 2004; Crossingham 2007):

1. The first stage is to select the data.
2. The second stage involves pre-processing the data to ensure that it is ready for analysis. This stage involves discretizing the data and removing unnecessary data (cleaning the data).
3. If reducts are considered, the third stage is to use the cleaned data to generate reducts. A *reduct* is the most concise way in which we can discern object classes. In other words, a reduct is the minimal subset of attributes that enables the same classification of elements of the universe as the whole set of attributes. To cope with inconsistencies, lower and upper approximations of decision classes are defined in this stage.
4. Stage four is where the rules are extracted or generated. The rules are usually determined based on condition attribute values. Once the rules are extracted, they can be presented in an '*if* CONDITION(S)-*then* DECISION' format.
5. The fifth and final stage involves testing the newly created rules on a test set. The accuracy must be noted and sent back into the optimization method used in step 2 and the process will continue until the optimum or highest accuracy is achieved.

The procedure for computing rough sets and extracting rules is given in Algorithm 8.1 (Crossingham 2007). Once the rules are extracted, they can be tested

Algorithm 8.1 Procedure to Generate a Rough Set Model

Input: Condition and Decision Attributes
Output: Certain and Possible Rules

Obtain the data set to be used;
Repeat
  for *conditional_attribute* ← 1 to *size_of_training_data* **do**
    Pre-process data to ensure that it is ready for analysis;
    Discretize the data according to the optimization technique;
    Compute the lower approximation, as defined in Eq. 8.5;
    Compute the upper approximation, as defined in Eq. 8.6;
    From the general rules, calculate plausibility measures for an object x belonging to set X, as defined by Eq. 8.13;
    Extract the *certain* rules from the lower approximation generated for each subset;
    Similarly, extract the *possible* rules from the upper approximation of each subset;
    Remove the generated rules for the purposes of testing on unseen data;
    Compute the classifier performance using the AUC;
  End
until Optimization technique termination condition;

using a set of testing data. The classification output is expressed as a decision value which lies between 0 and 1. The accuracy of the rough set is determined using the Area Under the receiver operating characteristic Curve (AUC).

8.5 Optimized Rough Sets

This section of the chapter uses the ant colony optimization technique to discretize the rough set partition. The general procedure pursued in this chapter is best expressed mathematically as follows (Marwala 2009; Marwala and Lagazio 2011):

$$y = f(x, RP) \tag{8.17}$$

Here x is the input data – in this chapter it is the dissolved gas analysis data, RP is the rough set partition and y is the accuracy obtained when the model is tested on unseen data. For this chapter, the aim of the optimization process was to identify the RP parameters so that the accuracy y was maximized. To this end, the ant colony optimization method was applied and is the subject of the next section (Mpanza 2011).

8.5.1 Ant Colony System

Ant Colony Optimization (ACO) is a meta-heuristic optimization technique for estimating solutions for combinatorial optimization problems (Blum 2005; Mpanza and Marwala 2011; Mpanza 2011). It is a type of the swarm intelligence group.

Swarm intelligence is the emergent collective intelligence of groups of simple agents. Ant colony system was first proposed by Dorigo (1992). It is inspired by the natural behavior of ants as they work collectively sharing information to identify the best possible route between their colony and the food source. This autonomy is achieved by ants depositing pheromone on their trail that acts as a signal for other ants. In the end, the path with the highest concentration of pheromone deposit is the optimal path.

ACO is different from other optimization tools in that it explicitly incorporates prior information on the structure of a good solution with *a posteriori* information on the good solution that was attained earlier (Yaseen and AL-Slamy 2008; Mpanza 2011; Mpanza and Marwala 2011). Different to the discretization techniques such as equal width or equal frequency types (Marwala and Lagazio 2011), ACO is a non-deterministic procedure. Non-deterministic procedures produce different solutions for different implementation because of their randomness (Thantulage 2009). ACO techniques have been applied in estimating solutions for the traveling salesmen problem (Dorigo and Gambardella 1997; Cheng and Mao 2007), the network routing problem (Di Caro and Dorigo 1997), flowshop scheduling (Tavares Neto and Godinho Filho 2011), seismic design (Kaveh et al. 2010), part-machine clustering (Xing et al. 2010a, b), for missing data estimation (Leke and Marwala 2006), feature selection (Chen et al. 2010b) and for attribute reduction (Ke et al. 2008). These applications have demonstrated that ACO is capable of estimating NP-hard problems.

8.5.2 Artificial Ant

An *artificial ant* is a computational agent describing the behavior of a real ant by iteratively constructing a solution for the problem at hand. The current state, *i*, is an incomplete solution and an ant follows the pheromone trail to move to the following state, *j*, and a more complete solution (Carbonaro and Maniezzo 2003).

8.5.3 Ant Colony Algorithm

An *ant colony algorithm* tries to solve a problem by constructing a candidate solution and using the current solution to drive the ant's path towards a high quality solution (Dorigo and Blum 2004; Mpanza 2011). The full procedure can be as follows (Mpanza 2011; Dorigo and Blum 2004):

1. Set the trail pheromone (τ) and the heuristic function (η)
2. Estimate the best possible path (T) based on the current τ and η,
3. Update τ and η based on the best selected path T,
4. Repeat step 1 until the termination condition is attained

Trail pheromone (τ_{ij}) is an *a posteriori* desirability of selecting the path between i and j while heuristic function (η_{ij}) is the attractiveness of selecting the path between i and j. Parameters τ_{ij} and η_{ij} are set to equal values to avoid biasing the solution search.

8.5.4 Ant Routing Table

An *ant routing table* is an action choice rule. An ant at node i uses this table to choose the best possible move to the next node j. The probability of ant k at node i choosing node j as the next node is given by (Mpanza 2011; Dorigo and Blum 2004):

$$P_j^k = \begin{cases} \dfrac{\tau_{ij}^{\alpha}\eta_{ij}^{\beta}}{\sum\limits_{m \in N_i^k}^{n} \tau_{ij}^{\alpha}\eta_{ij}^{\beta}} & \textit{if } j \in N_i \\ 0 & \textit{otherwise} \end{cases} \tag{8.18}$$

Here N_i is the set of nodes available from node i. Parameters η and τ are tuning variables. The solution identified by ant k is the path travelled T_k. The path is assessed by the objective function L_k.

8.5.5 Pheromone Update

The cost of the solution is identified by ant k in path T_k is L_k. This objective function is applied to update the pheromone trail for reinforcement (Mpanza 2011; Dorigo and Blum 2004):

$$\Delta\tau_{ij}^k = \begin{cases} \dfrac{1}{L_k} & \textit{if } (i, j) \in T_k \\ 0 & \textit{otherwise} \end{cases} \tag{8.19}$$

The effect of updating the pheromone in this manner is so that large objective functions result in small changes, whereas small objective functions result in significant changes. This is to bias future searches towards the solution with the lowest objective function. After all the ants have identified their solution, the pheromone is updated as follows (Mpanza 2011; Dorigo and Blum 2004):

$$\tau_{ij} = (1 - \rho)\,\tau_{ij} + \sum_{k=0}^{m} \tau_{ij}^k \tag{8.20}$$

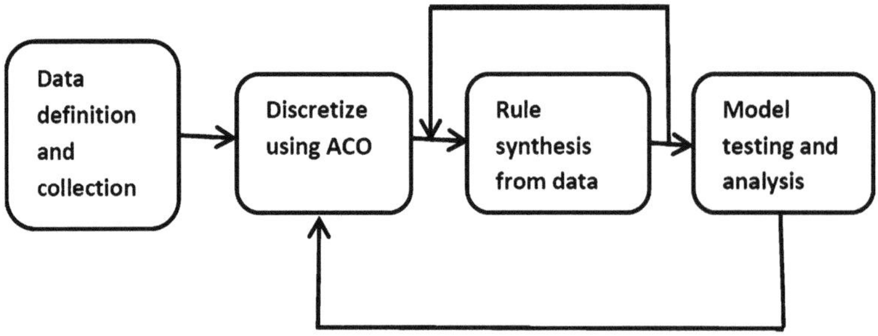

Fig. 8.1 Process followed in developing an ant colony optimization based rough set model (Mpanza 2011; Mpanza and Marwala 2011)

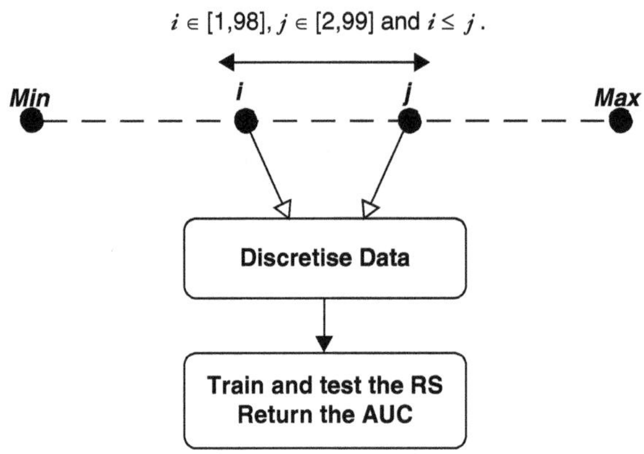

Fig. 8.2 The ant colony optimization discretization of a continuous variable. Here *RS* stands for rough sets and *AUC* stands for the area under the receiver operating characteristics curve (Mpanza 2011; Mpanza and Marwala 2011)

where ρ is the pheromone evaporation constant. This constant is used to reduce the pheromone on paths that are not sampled.

Previous studies have shown a number of techniques applied for data discretization but none of which applied ant-colony optimization. For this section, a rough set model was developed from data discretized using the ACO technique. The objective of this study was to:

1. Establish if ant colony optimization could be applied to discretize data for the detection of faults in bushings using rough set modeling.
2. Establish if ant colony optimization can contend with other discretization approaches. Figure 8.1 shows the process for developing an ant colony optimization model based on rough sets.

8.5.6 *Ant Colony Discretization*

Figure 8.2 shows the operation of ant colony optimization discretization (Mpanza 2011; Dorigo and Blum 2004). Discretization in ant colony optimization is

Algorithm 8.2 Ant Colony Optimized Rough Sets (Mpanza 2011)

```
Input: Data set to be discretized
Output: Discretized data
begin
    Initialize τ_ij and η_ij
    while Not termination condition do
        Data discretization using ant colony optimization
        foreach ant k do
            Compute τ_ij and η_ij
            Compute the probability of moving to state j;
            Choose the state with the highest probability;
            Append the chosen move to the ant's path T_k;
        end
        foreach ant move (i, j) do
            Compute Δτ_ij;
        end
        Compute the cost function L_k;
        Update trail matrix τ_ij;
        Save current best solution
    end
end
```

multivariate process and Fig. 8.2 shows a continuous variable being discretized in i and j, integer percentages of the range of a variable. To discretize a variable into three sections, two division points are needed. These division points ensure that ants estimate a solution by choosing an $(i; j)$ permutation in each variable that maximizes the testing accuracy of the model. Algorithm 8.2 demonstrates the ACO rough set process.

For this chapter the ACO discretization process was compared with the equal-frequency-bin (EFB) partitioning method as investigated by Crossingham (2007).

8.6 Application to Transformer Bushings

The technique was applied to a set of 2,000 bushings which were divided into training and testing bushings. The inputs were each discretized into three sections using the ACO and the EFB. Table 8.1 summarises the model's performance. EFB

Table 8.1 Comparison of the ACO and EFB (Mpanza 2011)

Method	Accuracy (%)	AUC	Number of rules
EFB	96.4	0.964	206
Ant colony	96.1	0.961	45

had an accuracy of 96.4% while the ACO had 96.1% which is approximately the same (Mpanza 2011). Nevertheless, the number of rules were substantially less in using the ACO than the EFB. Consequently, the ACO model has roughly the same degree of accuracy as the EFB, but is more transparent. The EFB surpasses the ACO in terms of computational time. However, this is barely a disadvantage for the ACO because training is a once-off process. Invoking Occam's razor, ACO is simpler and thus the better of the two classifiers.

8.7 Conclusion

This chapter presented a technique for discretizing input data for rough set modeling using ant colony optimization, a metaheuristic optimization method. The theories of rough set and ant colony optimization method were described. The presented technique was tested for the condition monitoring of transformer bushings. The ant colony optimization method was then compared to the equal frequency bin model and was observed to be better in transparency and equal in accuracy even though it was computationally expensive.

References

An YZ, Wen X, Li X, Xu Y, Long Y-K (2011) Morphological operator and rough set theory for fault line detection. In: 2011 Asia-Pacific Power and Energy Engineering Conference, APPEEC 2011 – Proceedings, Wuhan, art. no. 5748966

Azadeh A, Saberi M, Moghaddam RT, Javanmardi L (2011) An integrated data envelopment analysis – artificial neural network-rough set algorithm for assessment of personnel efficiency. Expert Syst Appl 38:1364–1373

Bazan J, Nguyen HS, Szczuka M (2004) A view on rough set concept approximations. Fundam Inform 59:107–118

Beynon M (2001) Reducts within the variable precision rough sets model: a further investigation. Eur J Oper Res 134:592–605

Bi Y, Anderson T, McClean S (2003) A rough set model with ontologies for discovering maximal association rules in document collections. Knowl Based Syst 16:243–251

Blum C (2005) Ant colony optimization: introduction and recent trends. Phys Life Rev 2(4): 353–373

Carbonaro A, Maniezzo V (2003) The ant colony optimization paradigm for combinatorial optimization. Adv Evol Comput, 1:539–557

Chen C, Shen J, Chen B, Shang CX, Wang YC (2010a) Building symptoms diagnosis criteria of traditional Chinese medical science treatment on the elderly's pneumonia by the rough set theory. In: Proceedings of the 29th Chinese control conference, Beijing, China pp 5268–5271

Chen Y, Miao D, Wang R (2010b) A rough set approach to feature selection based on ant colony optimization. Pattern Recognit Lett 31(3):226–233

Chen R, Zhang Z, Wu D, Zhang P, Zhang X, Wang Y, Shi Y (2011) Prediction of protein interaction hot spots using rough set-based multiple criteria linear programming. J Theor Biol 269:174–180

Cheng CB, Mao CP (2007) A modified ant colony system for solving the travelling salesman problem with time windows. Math Comput Model 46(9–10):1225–1235

Crossingham B (2007) Rough set partitioning using computational intelligence approach. MSc thesis, University of the Witwatersrand, Johannesburg

Crossingham B, Marwala T (2007) Using optimisation techniques to granulise rough set partitions. Comput Model Life Sci 952:248–257

Crossingham B, Marwala T (2008a) Using genetic algorithms to optimise rough set partition sizes for HIV data analysis. Stud Comput Intell 78:245–250

Crossingham B, Marwala T (2008b) Using optimisation techniques for discretizing rough set partitions. Int J Hybrid Intell Syst 5:219–236

Crossingham B, Marwala T, Lagazio M (2008) Optimised rough sets for modeling interstate conflict. In: Proceedings of the IEEE international conference on systems, man, and cybernetics, Singapore, pp 1198–1204

Crossingham B, Marwala T, Lagazio M (2009) Evolutionarily optimized rough set partitions. ICIC Expr Lett 3:241–246

Degang C, Wenxiu Z, Yeung D, Tsang ECC (2006) Rough approximations on a complete completely distributive lattice with applications to generalized rough sets. Info Sci 176:1829–1848

Di Caro G, Dorigo M (1997) Antnet: a mobile agent approach to adaptive routing. Techincal report IRIDIA/97-12. Universit Lubre de Bruxelles, Belgium

Dorigo M (1992) Optimization, learning and natural algorithms. PhD thesis, Politecnico di Milano, Milano

Dorigo M, Blum C (2004) Ant colony optimization. MIT Press, Cambridge

Dorigo M, Gambardella LM (1997) Ant colony system: a cooperative learning approach to the traveling salesman problem. IEEE Trans Evol Comput 1:53–66

Dubois D (1990) Rough fuzzy sets and fuzzy rough sets. Int J Gen Syst 17:191–209

Fayyad U, Irani K (1993) Multi-interval discretization of continuous valued attributes for classification learning. In: Proceedings of the 13th international joint conference on artificial intelligence, Los Alamos, CA, pp 1022–1027

Goh C, Law R (2003) Incorporating the rough sets theory into travel demand analysis. Tourism Manag 24:511–517

Gong J, Yang H, Zhong L (2010) Case-based reasoning based on rough set in rare-earth extraction process. In: Proceedings of the 29th Chinese control conference, Beijing, China, pp 70–1706

Grzymala-Busse JW (2004) Three approaches to missing attribute values – a rough set perspective. In: Proceedings of the IEEE 4th international conference on data mining, Brighton, UK, pp 57–64

Grzymala-Busse JW, Hu M (2001) A comparison of several approaches to missing attribute values in data mining. Lect Notes Artif Intell 205:378–385

Grzymala-Busse JW, Siddhaye S (2004) Rough set approaches to rule induction from incomplete data. In: Proceedings of the 10th international conference on infomation processing and management of uncertainty in knowledge-based systems, vol 2, Perugia, pp 923–930

Hoa NS, Son NH (2008) Rough set approach to approximation of concepts from taxonomy. http://logic.mimuw.edu.pl/publikacje/SonHoaKDO04.pdf. Last Accessed 9 July 2011

Huang C-C, Liang W-Y, Shian-Hua L, Tseng T-L, Chiang H-Y (2011) A rough set based approach to patent development with the consideration of resource allocation. Expert Syst Appl 38:1980–1992

Inuiguchi M, Miyajima T (2007) Rough set based rule induction from two decision tables. Eur J Oper Res 181:1540–1553

Jaafar AFB, Jais J, Hamid MHBHA, Rahman ZBA, Benaouda D (2006) Using rough set as a tool for knowledge discovery in DSS. In: Proceedings of the 4th international conference

on multimedia and information and communication technolgy in education, Seville, Spain, pp 1011–1015
Kaveh A, Farahmand Azar B, Hadidi A, Rezazadeh Sorochi F, Talatahari S (2010) Performance-based seismic design of steel frames using ant colony optimization. J Constr Steel Res 66(4):566–574
Ke L, Feng Z, Ren Z (2008) An efficient ant colony optimization approach to attribute reduction in rough set theory. Pattern Recognit Lett 29(9):1351–1357
Komorowski J, Pawlak Z, Polkowski L, Skowron A (1999) A rough set perspective on data and knowledge. In: Klösgen W, Zytkow JM, Klosgen W, Zyt J (eds) The handbook of data mining and knowledge discovery. Oxford University Press, New York
Kondo M (2006) On the structure of generalized rough sets. Info Sci 176:589–600
Leke BB, Marwala T (2006) Ant colony optimization for missing data estimation. In: Proceedings of the Pattern Recognition Association of South Africa, 2006, Parys, South Africa, pp 183–188
Li X, Li S, Xu Z (2005) Condition monitoring and fault diagnosis based on rough set theory. Yi Qi Yi Biao XueBao/Chin J Sci Instrum 26(suppl):781–783
Li L, Yang Z, He Z (2010) Research on intelligent fault diagnosis method based on rough set theory and fuzzy petri nets. Appl Mech Mater 26–28:77–82
Liao SH, Chen YJ, Chu PH (2010) Rough-set-based association rules applied to brand trust evaluation model. Lect Notes Comput Sci 6443:634–641
Lin CS, Tzeng GH, Chin YC (2011) Combined rough set theory and flow network graph to predict customer churn in credit card accounts. Expert Syst Appl 38:8–15
Machowski LA, Marwala T (2005) Using object oriented calculation process framework and neural networks for classification of image shapes. Int J Innov Comput, Info Control 1:609–623
Marwala T (2009) Computational intelligence for missing data imputation, estimation and management: knowledge optimization techniques. IGI Global Publications, New York
Marwala T, Crossingham B (2008) Neuro-rough models for modelling HIV. In: Proceedings of the IEEE international conference on systems, man, and cybernetics, Singapore, pp 3089–3095
Marwala T, Crossingham B (2009) Bayesian rough sets. ICIC Expr Lett 3:115–120
Marwala T, Lagazio M (2011) Militarized conflict modeling using computational intelligence techniques london. Springer, London
Mpanza LJ (2011) A rough set approach to bushings fault detection. University of Johannesburg Master of Engineering Thesis
Mpanza LJ, Marwala T (2011) Artificial neural network and rough set for HV bushings condition monitoring. In: Proceedings of the 15th IEEE international conference on intelligent engineering systems, Poprad, Slovakia
Mu Y, Xia H, Liu Y-K (2011) Fault diagnosis method for nuclear power plant based on decision tree and neighborhood rough sets. Yuanzineng Kexue Jishu/Atomic Energy Sci Technol 45(1):44–47
Nelwamondo FV (2008) Computational intelligence techniques for missing data imputation. PhD thesis, University of the Witwatersrand, Johannesburg
Ohrn A (1999) Discernibility and rough sets in medicine: tools and applications. unpublished PhD thesis, Norwegian University of Science and Technology
Ohrn A, Rowland T (2000) Rough sets: a knowledge discovery technique for multifactorial medical outcomes. Am J Phys Med Rehabil 79:100–108
Pang Q, Liu X, Zhang M (2010) Improved neural network based on rough set and application in fault line detection for distribution network. In: Proceedings of 2010 3rd international congress on image and signal processing, CISP 2010, 8, Yantai, China, art. no. 5646686, pp 3784–3788
Pawlak Z (1991) Rough sets – theoretical aspects of reasoning about data. Kluwer Academic Publishers, Dordrecht
Pawlak Z, Munakata T (1996) Rough control application of rough set theory to control. In: Proceedings of the 4th European congress on intelligence techniques and soft computing, Aachen, pp 209–218
Pawlak Z, Skowron A (2007) Rough sets and boolean reasoning. Info Sci 177:41–73

Pawlak Z, Wong SKM, Ziarko W (1988) Rough sets: probabilistic versus deterministic approach. Int J Man Mach Stud 29:81–95

Quafafou M (2000) α-RST: a generalization of rough set theory. Info Sci 124:301–316

Rowland T, Ohno-Machado L, Ohrn A (1998) Comparison of multiple prediction models for ambulation following spinal cord injury. In Chute 31:528–532

Sakthivel NR, Sugumaran V, Nair BB (2010) Comparison of decision tree-fuzzy and rough set-fuzzy methods for fault categorization of mono-block centrifugal pump. Mech Syst Signal Process 24(6):1887–1906

Salamó M, López-Sánchez M (2011) Rough set based approaches to feature selection for case-based reasoning classifiers. Pattern Recognit Lett 32:280–292

Shan N, Ziarko W (1995) Data-based acquisition and incremental modification of classification rules. Comput Intell 11:357–370

Shen L, Tay FEH, Qu L, Shen Y (2000) Fault diagnosis using rough sets theory. Comput Ind 43(1):61–72

Slezak D, Ziarko W (2005) The investigation of the Bayesian rough set model. Int J Approx Reasoning 40:81–91

Stefanowski J (1998) On rough set based approaches to induction of decision rules. In: Polkowski L, Skowron A (eds) Rough sets in knowledge discovery 1: Methodology and applications. Physica-Verlag, Heidelberg

Tavares Neto RF, Godinho Filho M (2011) An ant colony optimization approach to a permutational flowshop scheduling problem with outsourcing allowed. Comput Oper Res 38(9):1286–1293

Tay FEH, Shen L (2003) Fault diagnosis based on rough set theory. Eng Appl Artif Intell 16(1):39–43

Terlecki P, Walczak K (2007) On the relation between rough set reducts and jumping emerging patterns. Info Sci 177:74–83

Tettey T, Nelwamondo FV, Marwala T (2007) HIV data analysis via rule extraction using rough sets. In: Proceedings of the 11th IEEE international conference on intelligence engineering systems, Budapest, Hungary, pp 105–110

Thantulage GIF (2009) Ant colony optimization based simulation of 3d automatic hose /pipe routing. PhD thesis, School of Engineering and Design, Brunel University, London, UK

Wang J, Guo K, Wang S (2010) Rough set and tabu search based feature selection for credit scoring. Proced Comput Sci 1:2433–2440

Wang QH, Li JH (2004) A rough set-based fault ranking prototype system for fault diagnosis. Eng Appl Artif Intell 17(8):909–917

Wang W, Yang J, Jensen R, Liu X (2006) Rough set feature selection and rule induction for prediction of malignancy degree in brain glioma. Comput Methods Programs Biomed 83:147–156

Witlox F, Tindemans H (2004) The application of rough sets analysis in activity based modelling: opportunities and constraints. Expert Syst Appl 27:585–592

Wu W, Mi J, Zhang W (2003) Generalized fuzzy rough sets. Info Sci 151:263–282

Xiang X, Zhou J, Li C, Li Q, Luo Z (2009) Fault diagnosis based on walsh transform and rough sets. Mech Syst Signal Process 23(4):1313–1326

Xie F, Lin Y, Ren W (2011) Optimizing model for land use/land cover retrieval from remote sensing imagery based on variable precision rough sets. Ecol Model 222:232–240

Xing B, Gao WJ, Nelwamondo FV, Battle K, Marwala T (2010a) Ant colony optimization for automated storage and retrieval system. In: IEEE CEC 2010, Barcelona, Spain, pp 1133–1139

Xing B, Gao WJ, Nelwamondo FV, Battle K, Marwala T (2010b) Part-machine clustering: the comparison between adaptive resonance theory neural network and Ant Colony System 2010. Lect Notes Electrical Eng 67:747–755, Springer

Yan W, Liu W, Cheng Z, Kan J (2010) The prediction of soil moisture based on rough set-neural network model. In: Proceedings of the 29th Chinese control conference, Beijing, China, pp 2413–2415

Yang Y, John R (2006) Roughness bound in set-oriented rough set operations. In: Proceedings of the IEEE international conference on fuzzy systems, Vancouver, Canada, pp 1461–1468

Yaseen SG, AL-Slamy NMA (2008) Ant colony optimization. Int J Comput Sci Netw Secur 8(6):351–357

Yu Y, Han C (2010) Fault diagnosis of metro shield machine based on rough set & neural network. In: Proceedings – 3rd international conference on intelligent networks and intelligent systems, ICINIS 2010, Shenyang, China, art. no. 5693773, pp 588–591

Zhang L, Ji SM, Xie Y, Yuan QL, Zhang YD, Yao ZN (2008) Intelligent tool condition monitoring system based on rough sets and mathematical morphology. Appl Mech Mater 10–12:722–726

Zhang Y, Zhu J, Zhang Z-Y (2010) The research of reagent adding control in anionic reverse flotation process based on rough set theory. In: Proceedings of the 29th Chinese control conference, Beijing, China, pp 3487–3491

Zhao X, Zuo MJ, Patel T (2010) Application of fuzzy preference based rough set model to condition monitoring. Lect Notes Comput Sci (including subseries Lecture Notes in Artificial Intelligence and Lecture Notes in Bioinformatics) 6086 LNAI:688–697

Zhao Y, Yao Y, Luo F (2007) Data analysis based on discernibility and indiscernibility. Info Sci 177:4959–4976

Zou Z, Tseng T-L, Sohn H, Song G, Gutierrez R (2011) A rough set based approach to distributor selection in supply chain management. Expert Syst Appl 38:106–115

Chapter 9
Condition Monitoring with Incomplete Information

9.1 Introduction

Neural networks have been applied in together with vibration data to identify faults in structures (Doebling et al. 1996; Marwala 2001, 2003, 2004; Zang and Imregun 2001; Waszczyszyn and Ziemianski 2001; Marwala and Chakraverty 2006). Neural networks estimate functions of arbitrary complexity using training data. Supervised neural networks are applied to characterize a mapping from an input vector onto an output vector, whereas unsupervised networks are applied to classify the data without prior knowledge of the classes involved. One of the the most used neural network architectures is the multi-layer perceptron (MLP) which is trained using the back-propagation technique (Marwala 2000; Bishop 1995). Another network type is the radial basis function or RBF (Bishop 1995). Both the MLP and RBF have been applied for fault identification in structures but it has been observed that MLP generally performs better than RBF (Marwala 2000). This is because the RBF generally requires the application of the pseudo-inverse of a matrix for training, which is normally singular while the MLP applies optimization approaches that are stable (Marwala 2000).

Levin and Lieven (1998) used a RBF neural network and modal properties to identify errors in a finite element model of a cantilevered beam. The technique was found to give a good identification of the faults even with a limited number of experimentally measured degrees of freedom and modes.

Wu et al. (1992) applied an MLP neural network to identify the damage in a model of a three-story building. Damage was modeled by reducing member stiffness between 50% and 75%. The input to the neural network was a Fourier transform of the acceleration data and the output was the level of damage in each member. The network could diagnose damage within 25% accuracy.

Lopes et al. (2000) applied impedance techniques and neural networks for structural health monitoring. Marwala (2001) applied a probabilistic committee of neural networks to classify faults in a population of nominally identical cylindrical shells. The probabilistic neural networks were trained using the hybrid Monte Carlo

T. Marwala, *Condition Monitoring Using Computational Intelligence Methods: Applications in Mechanical and Electrical Systems*, DOI 10.1007/978-1-4471-2380-4_9,

(Neal 1993) and an accuracy of 95% was observed in classifying the eight-classes fault cases. Chen et al. (2003) applied neural networks and response-only data for the fault diagnosis of structures.

Atalla and Inman (1998) trained a RBF neural network using frequency response functions to identify faults in structures. Marwala and Hunt (1999) applied multi-layer perceptron neural networks and finite element models to identify faults in a cantilevered beam. Atalla and Inman (1998) trained a RBF neural network using frequency response functions to identify faults in structures. Suresh et al. (2004) used a modular neural network approach to identify crack location in a cantilever beam and elsewhere Reddy and Ganguli (2003) used radial basis function neural networks for a helicopter rotor blade. Pawar and Ganguli (2003) used a genetic fuzzy system for damage detection in beams and helicopter rotor blades.

When these neural networks are applied in a real life condition, one of the key problems faced is the issue of sensor failure. If one of the sensors fails, then the neural network is not capable of making a decision because it only works with a complete input set. What is normally done is to use the average value of that sensor calculated over some defined period in the past and hope that the next time around the sensor will be available.

In the literature there is no method presented thus far that takes account of the *absence* of entries of inputs to the neural networks for fault classification in structures. It must be noted, nevertheless, that the issue of estimating missing data has been applied in other areas in mechanical systems such as validating the gas-path sensor data (Lu and Hsu 2002). Consequently this chapter contributes to the field of structural mechanics a technique that has been applied in gas dynamics.

This chapter explains a technique for approximating missing entries in the database that is based on auto-associative models (Kramer 1992) combined with genetic algorithm for data estimation and subsequently fault identification in structural mechanics (Marwala and Chakraverty 2006). This technique was tested on a classification of faults in a population of nominally cylindrical shells.

9.2 Mathematical Background

The mathematical background to neural networks and auto-associative networks with Missing Data is explained in this section.

9.2.1 Neural Networks

This chapter we applied neural networks to construct auto-associative neural networks. These are networks with inputs and output being the same (Kramer 1992; Marwala and Chakraverty 2006). There are different types of neural network topologies but this chapter focuses on the MLP. The MLP network implemented in this chapter contains a hyperbolic tangent basis function in the hidden units and linear basis functions in the output units (Bishop 1995). The relationship between the output y and input x be written as follows (Bishop 1995):

$$y_k = \sum_{j=0}^{M} w_{kj}^{(2)} \tanh\left(\sum_{i=0}^{d} w_{ji}^{(1)} x_i\right) \tag{9.1}$$

where $w_{ji}^{(1)}$ and $w_{kj}^{(2)}$ indicate weights in the first and second layer, respectively, going from input i to hidden unit j, M is the number of hidden units, and d is the number of output units.

The MLP model can take into account the intrinsic dimensionality of the data. Models of this nature can estimate any continuous function to arbitrary accuracy if the number of hidden units M is adequately large. A training of the neural network identifies the weights in Eq. 9.1. A cost or objective function must be selected to identify the weights in Eq. 9.1. A *cost function* is a mathematical representation of the overall objective of the problem. In this chapter, the main objective is to construct the cost function that identifies a set of neural network weights given the measured data. If the training set $D = \{x_k, t_k\}_{k=1}^{N}$ is used, and assuming that the targets y are sampled independently given the inputs x_k and the weight parameters, w_{kj}, then the cost function, E, may be written using the sum-of-squares of errors cost function (Bishop 1995):

$$E = \sum_{n=1}^{N} \sum_{k=1}^{K} \{t_{nk} - y_{nk}\}^2 \tag{9.2}$$

here t is the target data, N is the number of training examples and K is the number of outputs.

Before network training is performed, the network architecture needs to be created by selecting the number of hidden units, M. If M is too small, the neural network will be inadequately flexible and will give a poor generalization of the data because of high bias. Contrariwise, if M is too large, the neural network will be unnecessarily flexible and will give a poor generalization due to the phenomenon known as *over-fitting* caused by high variance. In this study to minimize the equation, the scaled conjugate gradient technique was applied in conjunction with back-propagation (Møller 1993; Bishop 1995). The scaled conjugate gradient technique is an optimization method that is based on the conjugate gradient technique but uses optimized mathematical expressions to reduce the computational intensity of the conjugate gradient method. It must however be noted that there is no material difference in accuracy of the results of the scaled conjugate gradient, conjugate gradient and other gradient based optimization methods. The only difference between these methods is the computational efficiency and so the scaled conjugate gradient method was selected because of its computational efficiency.

9.2.2 Auto-Associative Networks with Missing Data

Auto-associative networks are models where the network is trained to recall its inputs. This means that whenever an input is presented to the network the output is the predicted input. These networks have been used in a number of applications including novelty detection, feature selection and data compression (Hines et al. 1998; Kramer 1991; Upadhyaya and Eryurek 1992; Jensen et al. 2001; Reed and Marks II 1999).

There has been an increased interest in handling the missing data problem by estimation or imputation (Nelwamondo and Marwala 2007; Abdella and Marwala 2006). The combination of the auto-associative neural network and the genetic algorithm (GA) has been demonstrated to be a successful method to estimate missing data (Nelwamondo 2008; Abdella and Marwala 2006). The efficient and effective estimation of missing data relies on the extraction and storage of the relationships or correlations between the variables that make up the dataset (Nelwamondo 2008). Auto-associative neural networks allow this to be conducted well (Kramer 1991) Nevertheless, other methods such as a standard principal component analysis (PCA) can also be used successfully.

Other applications of auto-associative network includes its use in structural damage detection (Zhou et al. 2011), in autonomous single-pass end-member approximation (Ritter et al. 2009), in spoken cued recall (de Zubicaray et al. 2007) and in fault identification in rotating machinery (Sanz et al. 2007). Also, Amiri et al. (2008) analyzed the dynamical behaviour of a feedback auto-associative memory.

It must be borne in mind that on applying auto-associative neural networks for data compression, the network has fewer nodes in the hidden layer. However, it must be noted that for missing data estimation it is crucial that the network be as accurate as possible and that this accuracy is not necessarily realized through few hidden nodes as is the case when these networks are used for data compression. It is consequently vital that some process for identifying the optimal architecture must be used. By using Eq. 9.1 in an auto-associative memory network, as shown in Fig. 9.1, an auto-associative memory network may be formulated by setting the input x to be equal to the output y. Equation 9.1 may thus be re-written in simplified form as (Marwala 2009; Abdella and Marwala 2006):

$$\{y\} = f(\{w\}, \{x\}) \tag{9.3}$$

Here $\{y\}$ is the output vector, $\{x\}$ is the input vector, f is a function, and $\{w\}$ is the mapping weight vector. Given the fact that $\{x\} = \{y\}$, Eq. 9.3 may thus be re-written as follows (Marwala 2009; Abdella and Marwala 2006):

$$\{x\} = f(\{w\}, \{x\}) \tag{9.4}$$

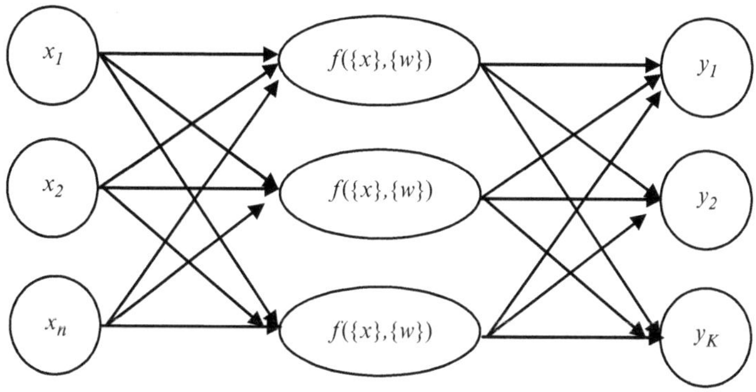

Fig. 9.1 An auto-associative MLP network having two layers of adaptive weights

For a perfectly mapped system, Eq. 9.4 holds. Nevertheless, for a realistic mapping there will be some error, and thus Eq. 9.4 may be re-written as (Marwala 2009; Abdella and Marwala 2006):

$$\{e\} = \{x\} - f(\{w\}, \{x\}) \tag{9.5}$$

The sum of squares of both the left hand side and the right hand side of Eq. 9.5 will give (Marwala 2009; Abdella and Marwala 2006):

$$E = \sum_{i=1}^{c} (\{x\} - f(\{w\}, \{x\}))^2 \tag{9.6}$$

Here c is the size of the input vector. For a situation when not all the inputs are known, the input data may be divided into known x_{kw} and unknown components x_u and thus Eq. 9.6 may be written as follows (Marwala 2009; Abdella and Marwala 2006):

$$E = \sum_{i=1}^{c} \left(\left\{ \begin{array}{c} x_u \\ x_{kw} \end{array} \right\} - f\left(\{w\}, \left\{ \begin{array}{c} x_u \\ x_{kw} \end{array} \right\} \right) \right)^2 \tag{9.7}$$

From Eq. 9.7, the unknown component data x_u is estimated from the known component x_{kw} by minimizing the error in Eq. 9.7. It is essential that a global minimum error be achieved because a local minimum error results from the incorrect estimation of the unknown component x_u. For this chapter a global optimum technique, genetic algorithm, was applied to identify the global optimum solution (Holland 1975). The next section thus explains the genetic algorithm.

9.3 Genetic Algorithms (GA)

The condition monitoring procedure presented in this chapter used a genetic algorithm to estimate the missing data by minimizing Eq. 9.7. Different to various optimization procedures, a genetic algorithm technique has a higher probability of converging to a global optimal solution than does a gradient-based method (Marwala 2010). A *genetic algorithm* is a population-based, probabilistic method that operates to find a solution to a problem from a population of possible solutions (Goldberg 1989; Holland 1975; Kubalík and Lazanský 1999; Marwala 2009). It is applied to identify approximate solutions to difficult problems through the analogy of the principles of evolutionary biology to computer science (Michalewicz 1996; Mitchell 1996; Forrest 1996; Vose 1999; Tettey and Marwala 2006). It was inspired by Darwin's theory of evolution where members of the population compete to survive and reproduce whereas the weaker are eliminated from the population. Every individual is allocated a fitness value according to how well it satisfies the objective of solving the problem. New and more evolutionary-fit individual solutions are produced during a cycle of generations, where selection and recombination operations take place, similar to how gene transfer happens to the current individuals. This continues until a termination condition is satisfied, after which the best individual thus far is considered to be the estimation for missing data. This chapter describes the application of a genetic algorithm to optimize Eq. 9.7.

Applications of genetic algorithm include those in structures (Marwala 2002), in helicopter rotor-blade design (Akula and Ganguli 2003), in artificial boundary conditions (Tu and Lu 2008) and for material model parameter identification for low-cycle fatigue (Franulović et al. 2009). Other recent applications of GA include Balamurugan et al. (2008) who evaluated the performance of a two-stage adaptive genetic algorithm, enhanced with island and adaptive features for structural topology optimization. Elsewhere, Kwak and Kim (2009) used a hybrid genetic algorithm, improved by a direct search for optimum design of reinforced concrete frames. Canyurt et al. (2008) approximated the strength of a laser hybrid welded joint using a genetic-algorithm technique.

Perera et al. (2009) used a GA to evaluate the performance of a multi-criteria damage-identification system. Almeida and Awruch (2009) applied a GA to optimally design composite laminated structures. Their GA was adapted with particular operators and variables codification for the definite class of composite laminated structures.

Li and Du (2012) applied a method for handling the inequality constraint in Gas, using a boundary simulation method. Elsewhere, Gladwin et al. (2011) applied GA for hardware-in-the-loop experimentation. Mosalman Yazdi and Ramli Sulong (2001) applied a GA to the optimization of Off-Centre bracing, while Balin (2011) applied a GA in non-identical parallel machine scheduling, and Musharavati and Hamoud (2011) applied a modified GA for manufacturing process planning in manufacturing lines making multiple parts.

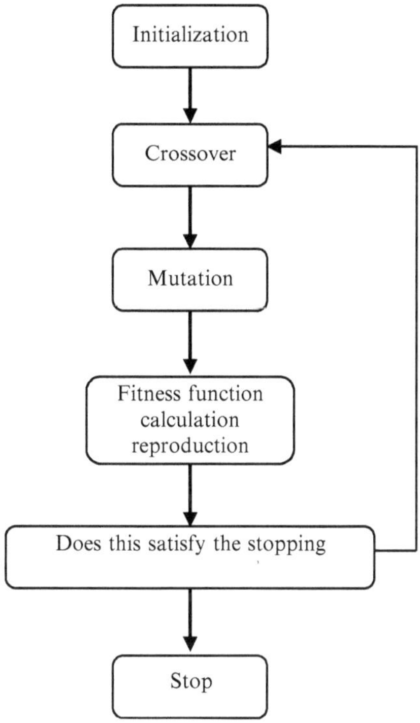

Fig. 9.2 Flow chart of the genetic algorithm method

Additional applications of a genetic algorithm for optimization structures include Paluch et al. (2008) as well as Roy and Chakraborty (2009). In addition, GA has also been proven to be successful in a variety of applications including:

- finite-element analysis (Marwala 2003, 2010);
- selecting optimal neural-network architecture (Arifovic and Gençay 2001);
- training hybrid fuzzy neural networks (Oh and Pedrycz 2006);
- solving job-scheduling problems (Park et al. 2003);
- remote sensing (Stern et al. 2006);
- missing-data estimation (Abdella and Marwala 2006; Marwala 2009); and
- combinatorial optimization (Zhang and Ishikawa 2004).

Additionally, GA has been proven to be successful in complex optimization problems such as wire-routing, scheduling, adaptive control, game-playing, cognitive modeling, transportation problems, traveling salesman problems, optimal control problems and database-query optimization (Pendharkar and Rodger 1999; Marwala et al. 2001; Marwala and Chakraverty 2006; Marwala 2007; Crossingham and Marwala 2007; Hulley and Marwala 2007). The MATLAB® implementation of a GA as described in Houck et al. (1995) was used as the GA in this chapter. To apply a GA, the following steps are followed as shown in Fig. 9.2: initialization, crossover, mutation, selection, reproduction, and termination.

In this chapter, the GA viewed learning as a competition between populations of evolving candidate problem solutions. A fitness function, which in this chapter is represented by Eq. 9.4, evaluates each solution to decide whether it will contribute to the next generation of solutions. Through operations analogous to gene transfer in sexual reproduction, the algorithm creates a new population of candidate solutions (Goldberg 1989). The three most important aspects of using a genetic algorithm are the:

- definition of the objective function;
- implementation of the genetic representation; and
- implementation of the genetic operators.

The details of genetic algorithms are illustrated in Fig. 9.2.

9.3.1 Initialization

In the beginning, a large number of possible individual solutions are randomly generated to form an initial population. This initial population is sampled so that it covers a good representation of the updating solution space. Within the context of this chapter, the size of the population should depend on the nature of the problem.

9.3.2 Crossover

The crossover operator fuses genetic information in the population by cutting pairs of chromosomes at random points along their length and swapping the cut sections over. This has a potential for assembling successful operators (Gwiazda 2006).

Crossover occurs with a certain probability. In many natural systems, the probability of crossover occurring is higher than the probability of mutation occurring. An example is a simple crossover technique (Banzhaf et al. 1998; Goldberg 1989).

For simple crossover, one crossover point is chosen, a binary string from the beginning of a chromosome to the crossover point is copied from one parent, and the rest is copied from the second parent. For instance, if two chromosomes in binary space $a = 11001011$ and $b = 11011111$ undertake a one-point crossover at the midpoint, then the resulting offspring is $c = 11001111$. For arithmetic crossover, a mathematical operator is performed to make an offspring. For example, an AND operator can be executed on $a = 11001011$ and $b = 11011111$ to create an offspring 11001011.

9.3.3 Mutation

The *mutation* operator chooses a binary digit in the chromosomes at random and inverts it. This has a potential of adding new information to the population, and in so doing avoids the GA simulation from being trapped in a local optimum solution. Mutation takes place with a certain probability. In many natural systems, the probability of mutation is low (*i.e.*, less than 1%). For this chapter, binary mutation was applied (Goldberg 1989). When binary mutation is applied, a number written in binary form was selected and one bit value was inverted. For instance: the chromosome 11001011 may become the chromosome 11000011.

Non uniform mutation operates by increasing the probability of mutation such that it will approximate 0 as the generation number increases adequately. It avoids the population from stagnating in the early stages of the evolution process, and then allows the procedure to improve the solution in the end stages of the evolution.

9.3.4 Selection

For each generation, a selection of the proportion of the existing population is selected to breed a new population. This selection is accomplished by using the fitness-based procedure, where solutions that are fitter, as measured by Eq. 9.7, are given a higher probability of being chosen. Some selection approaches rank the fitness of each solution and select the best solutions, whereas other techniques rank a randomly selected sample of the population for computational efficiency.

Numerous selection functions have a tendency to be stochastic in nature and are therefore designed such that a selection procedure is performed on a small proportion of less fit solutions. This ensures that diversity of the population of possible solutions is preserved at a high level and, consequently, avoids convergence on poor and incorrect solutions. There are numerous selection techniques. These include roulette-wheel selection (Mohamed et al. 2008).

Roulette-wheel selection is a genetic operator applied for selecting potentially valuable solutions in a GA optimization process. In this technique, each possible technique is assigned the fitness function that is applied to map the probability of selection with each individual solution. Supposing the fitness f_i is of individual i in the population then the probability that this individual is selected is:

$$p_i = f_i \Big/ \sum_{j=1}^{N} f_j \tag{9.8}$$

Here N is the total population size.

This procedure guarantees that candidate solutions with a higher fitness have a lower probability so that they may eliminate those with a lower fitness. In the same way, solutions with low fitness have a low probability of surviving the selection process. The advantage of this step is that although a solution may have low fitness, it may still have some components that may be beneficial in the future.

The process explained results in the following generation of a population of solutions that is different from the parent generation and that has an average fitness that is higher than the preceding generation.

Algorithm 9.1 An Algorithm for Implementing a Genetic Algorithm

1. Choose the initial population
2. Calculate the fitness of each chromosome in the population using Eq. 9.7
3. Repeat
   a. Select chromosomes with higher fitness to reproduce
   b. Produce a new population using crossover and mutation to create offsprings
   c. Calculate the fitness of each offspring
   d. Subsitute the low fitness section of the population with offspring
4. Repeat until termination

9.3.5 Termination

The procedure explained is recurred until a termination criterion has been realized, either because a desired solution that meets the objective function in Eq. 9.7 was identified or because a stated number of generations has been achieved or the solution's fitness has converged (or any combination of these).

The procedure explained can be expressed in pseudo-code, as described in Algorithm 9.1 (Goldberg 1989; Marwala 2010). For instance, for a GA representation, a choice has to be made between a binary and a floating-point representation. For the initialization procedure, a choice has to be made for the population size.

9.4 Missing Entry Methodology

The *missing data methodology* applied in this chapter combined the auto-associative neural networks and with the optimization technique, *viz.* the genetic algorithm. The technique is shown in Fig. 9.3. It was applied by determining the number of missing data and calling this number N. Then, the missing entry objective function with N variables was constructed using Eq. 9.7. A Genetic algorithm was then applied to minimize the missing entry objective function and the optimum solution is the approximated values of the missing variables. The missing entry objective function in the equation can also be solved using a gradient-based method because the gradient of the error function can be calculated simply by using back-propagation. Nevertheless, the gradient-based techniques are not global procedures and so were not used for this chapter although they can be used to adjust the solution given by the genetic algorithm. Nonetheless, it has been observed that fine-tuning the genetic algorithm solution does not offer any advantage in the estimation of the missing entries nor in the accuracy of fault classification (Marwala and Chakraverty 2006).

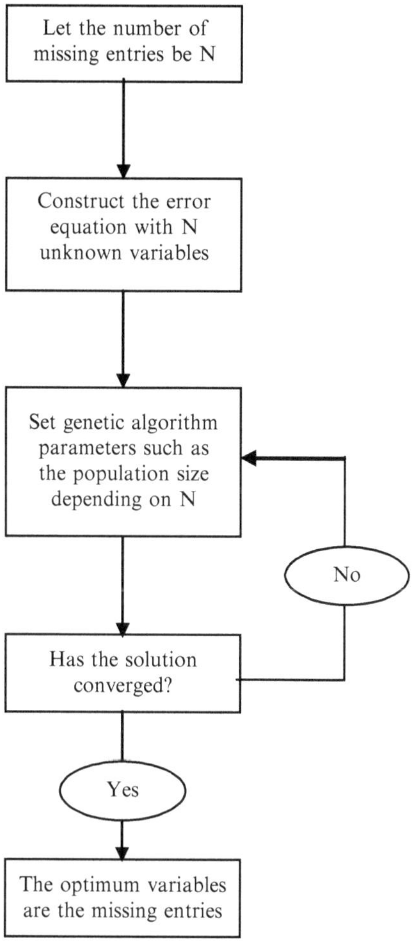

Fig. 9.3 A diagram indicating the implementation of the missing data estimator

9.5 Dynamics

In this chapter, modal properties *i.e.,* natural frequencies and mode shapes wee applied for the fault classification of the population of cylinders which were described in Chap. 2. Consequently these parameters are surveyed only briefly in this section. Modal properties are related to the physical properties of the structure. All elastic structures may be defined in terms of their distributed mass, damping and stiffness matrices in the time domain through the following expression (Ewins 1995; Marwala 2010):

$$[M]\{X''\} + [C]\{X'\} + [K]\}\{X\} = \{F\} \tag{9.9}$$

Here $[M]$, $[C]$ and $[K]$ are the mass, damping and stiffness matrices respectively, and $\{X\}$, $\{X'\}$ and $\{X''\}$ are the displacement, velocity and acceleration vectors, respectively. Finally $\{F\}$ is the applied force vector. If Eq. 9.9 is transformed into the modal domain to form an eigenvalue equation for the ith mode, then (Ewins 1995; Marwala 2010):

$$(-\overline{\omega}_i^2[M] + j\overline{\omega_i}[C] + [K])\{\overline{\varphi}\}_i = \{0\} \tag{9.10}$$

where $j=\sqrt{-1}$, $\overline{\omega}_i$ is the ith complex eigenvalue, with its imaginary part corresponding to the natural frequency ω_i, $\{0\}$ is the null vector, and $\{\overline{\varphi}\}_i$ is the ith complex mode shape vector with the real part corresponding to the normalized mode shape $\{\phi\}_i$. From Eq. 9.10 it may be assumed that changes in the mass and stiffness matrices cause changes in the modal properties of the structure. Consequently, the modal properties can be identified through the identification of the correct mass and stiffness matrices.

9.6 Example: A Cylindrical Structure

In this section the technique presented in this chapter was experimentally validated. The experiment was performed on a population of cylinders, which were supported by inserting a sponge rested on bubble-wrap plastic sheet, to simulate a 'free-free' environment. The particulars of this may be obtained in Marwala (2001) and was described in Chap. 2.

As described by Marwala (2001), each cylinder was divided into three equal substructures and holes 10–15 mm in diameter were introduced at the centers of the substructures to simulate faults. For one cylinder, the first type of fault was a zero-fault scenario. This type of fault was given the identity [0 0 0], indicating that there were no faults in any of the three substructures. The second type of fault was a one-fault-scenario, where a hole might be located in any of the three substructures. The three possible one-fault-scenarios were [1 0 0], [0 1 0], and [0 0 1] indicating one hole in substructures 1, 2, or 3 respectively. The third type of fault was a two-fault scenario, where a hole was located in two of the three substructures. The three possible two-fault-scenarios were [1 1 0], [1 0 1], and [0 1 1]. The final type of fault was a three-fault-scenario, where a hole was located in all three substructures, and the identity of this fault was [1 1 1]. There were eight different types of fault-cases considered (including [0 0 0]).

Because the zero-fault scenarios and the three-fault scenarios are over-represented, 12 cylinders were picked at random and additional one- and two-fault cases were measured after increasing the magnitude of the holes. This was done before the next fault case was introduced to the cylinders. The reason why zero-fault and three-fault scenarios are over-represented was because all cylinders tested give these fault-cases, whereas not all cylinders tested give all 3 one-fault and 3 two-fault cases. Only a few fault-cases were selected because of the limited computational storage space available. For each fault-case, acceleration and impulse measurements were taken. The types of faults that were introduced (*i.e.,* drilled holes) do not influence damping.

Each cylinder was measured three times under different directions by changing the orientation of a rectangular sponge inserted inside the cylinder. The number of sets of measurements taken for undamaged population was 60 (20 cylinders × 3 different directions).

The impulse and response data were processed using the Fast Fourier Transform to convert the time domain impulse history and response data into the frequency domain. The data in the frequency domain was used to calculate the FRFs. The FRF results from an ensemble of 20 undamaged cylinders indicated that the measurements were generally repeatable at low frequencies and were not repeatable at high frequencies. Axi-symmetric structures such as cylinders have repeated modes due to their symmetry (Royston et al. 2000). In the work for this chapter, the presence of an accelerometer and the imperfection of cylinders destroy the axi-symmetry of the structures. Therefore, the problem of repeated natural frequencies was neatly avoided thereby making the process of modal analysis easier (Maia and Silva 1997). The problem of uncertainty of high frequencies was avoided by only using frequencies under 4,000 Hz.

From the measured data, 10 parameters were selected using the principal component analysis described in Chap. 2. An auto-associative network with 10 inputs and 10 outputs was constructed and several numbers of hidden units were used as shown in Fig. 9.4 (Marwala and Chakraverty 2006). As shown in this figure, it was found that 10 hidden units was the optimal network that gives the best prediction of the input data. However it must be noted that it is generally assumed that the best auto-associative network is the one that has the lowest possible number of hidden units. However, in this study, *i.e.,* the case of missing data estimation, this factor was not taken for granted and it is recommended that a separate study, like the one conducted here, be used to determine the optimal auto-associative network. This is because for missing data estimation it was found that the success of the procedure was determined by how accurate the networks were and the accuracy did not necessarily only occur when the size of hidden nodes was small. As indicated before, the auto-associative network was trained using the scaled conjugate method.

The first experiment consisted of cases where one of the input to the neural network was assumed to be unknown and then estimated using a genetic algorithm method. To apply a genetic algorithm, an arithmetic cross-over, non-uniform mutation and normalized geometric selection were used. On applying the arithmetic cross-over, a number of parameters had to be selected. These were bounds and the probability of cross-over. The bounds were determined from the maxima and minima of historical values of the particular data point, while the probability of cross-over was selected to be 0.75 as suggested by Holland (1975). On implementing the mutation the parameters that needed to be chosen were the bounds, and these were chosen as for cross-over, and the probability of mutation, was chosen

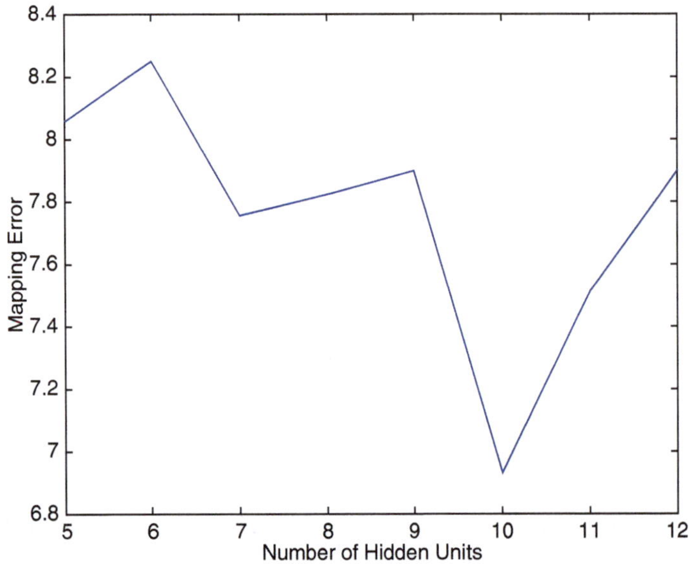

Fig. 9.4 The prediction error versus the number of hidden nodes

to be 0.0333 as recommended by Goldberg (1989). The Genetic algorithm had a population of 20 and was run for 25 generations.

The presented method for the case of one missing data per input set, estimated the missing value to the accuracy of 93%. When the method was tested for the case with two missing data per input set, the accuracy of the estimated values was 91%. The estimated values together with the accurate values are also indicated in Fig. 9.5 (Marwala and Chakraverty 2006). This figure illustrates that the missing data estimator gave results that are consistent and accurate. In fact the data in this figure shows that the correlation between the estimated data and the correct data was 0.9.

In many cases the estimated values were intended for a particular reason. In this chapter they were intended to fulfill the goal of fault classification in a population of cylinders. The estimated values were, therefore, used for the classification of faults in a population of cylindrical shells and the fault classification accuracy of 94% was observed for a one-missing-entry case and 91% for the two-missing-entry case. When the complete database was used, a fault classification accuracy of 96% was achieved.

The sources of errors in the experiment were measurement errors, modal analysis and neural network training. To minimize these errors, reliable instruments were used for measuring data, reliable software was used for signal processing and modal analysis and standard procedures were used for training, generalization and testing of neural networks. The impact of these errors on the quality of results was to such a small degree that it did not compromise the quality of the results.

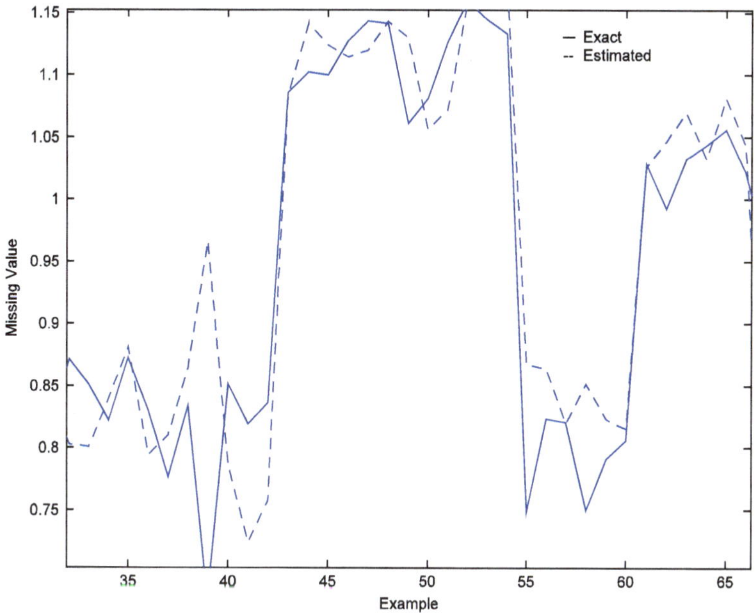

Fig. 9.5 The measured and estimated missing value

9.7 Conclusion

In this chapter, a technique based on auto-associative neural networks and genetic algorithms was presented to approximate missing entries in data. This technique was tested on a population of cylindrical shells. The technique could approximate single-missing-entries to an accuracy of 93% and two-missing-entries to an accuracy of 91%. Furthermore, a fault classification accuracy of 94% was obtained for single-missing-entry cases and 91% for two-missing-entry cases whereas the full database set gave a classification accuracy of 96%.

References

Abdella M, Marwala T (2006) The use of a genetic algorithm and neural networks to approximate missing data in database. Comput Inform 24:1001–1013

Akula VR, Ganguli R (2003) Finite element model updating for helicopter rotor blade using a genetic algorithm. AIAA J. doi:10.2514/2.1983

Almeida FS, Awruch AM (2009) Design optimization of composite laminated structures using a genetic algorithm and finite element analysis. Compos Struct 88:443–454

Amiri M, Saeb S, Yazdanpanah MJ, Seyyedsalehi SA (2008) Analysis of the dynamical behavior of a feedback auto-associative memory. Neurocomputing 71:486–494

Arifovic J, Gençay R (2001) Using a genetic algorithms to select architecture of a feedforward artificial neural network. Phys A: Stat Mech Appl 289:574–594
Atalla MJ, Inman DJ (1998) On model updating using neural networks. Mech Syst Signal Process 12:135–161
Balamurugan R, Ramakrishnan CV, Singh N (2008) Performance evaluation of a two stage adaptive genetic algorithm (TSAGA) in structural topology optimization. Appl Soft Comput 8:1607–1624
Balin S (2011) Non-identical parallel machine scheduling using genetic algorithm. Expert Syst Appl 38:6814–6821
Banzhaf W, Nordin P, Keller R, Francone F (1998) Genetic programming – an introduction: on the automatic evolution of computer programs and its applications. Morgan Kaufmann Publishers, San Francisco
Bishop CM (1995) Neural networks for pattern recognition. Oxford University Press, Oxford
Canyurt OE, Kim HR, Lee KY (2008) Estimation of laser hybrid welded joint strength by using a genetic algorithm approach. Mech Mater 40:825–831
Chen Q, Chan YW, Worden K (2003) Structural fault diagnosis and isolation using neural networks based on response only data. Comput Struct 81:2165–2172
Crossingham B, Marwala T (2007) Using a genetic algorithms to optimise rough set partition sizes for HIV data analysis. Stud Comput Intell 78:245–250
de Zubicaray G, McMahon K, Eastburn M, Pringle AJ, Lorenz L, Humphreys MS (2007) Support for an auto-associative model of spoken cued recall: evidence from fMRI. Neuropsychologia 45:824–835
Doebling SW, Farrar CR, Prime MB, Shevitz DW (1996) Damage identification and health monitoring of structural and mechanical systems from changes in their vibration characteristics: a literature review. Los Alamos National Laboratory report LA-13070-MS
Ewins DJ (1995) Modal testing: theory and practice. Research Studies Press, Letchworth
Forrest S (1996) Genetic algorithms. ACM Comput Surv 28:77–80
Franulović M, Basan R, Prebil I (2009) A genetic algorithm in material model parameters' identification for low-cycle fatigue. Comput Mater Sci 45:505–510
Gladwin D, Stewart P, Stewart J (2011) A controlled migration genetic algorithm operator for hardware-in-the-loop experimentation. Eng Appl Art Intell 24:586–594
Goldberg DE (1989) Genetic algorithms in search, optimization and machine learning. Addison-Wesley, Reading
Gwiazda TD (2006) A genetic algorithms reference Vol.1 Cross-over for single-objective numerical optimization problems. Adobe eBook, Lomianki
Hines JW, Uhrig RE, Wrest DJ (1998) Use of autoassociative neural networks for signal validation. J Intell Robot Syst 21:143–154
Holland J (1975) Adaptation in natural and artificial systems. University of Michigan Press, Ann Arbor
Houck CR, Joines JA, Kay MG (1995) A genetic algorithm for function optimisation: a MATLAB implementation. Technical Report NCSU-IE TR 95–09, North Carolina State University, Raleigh
Hulley G, Marwala T (2007) A genetic algorithm based incremental learning for optimal weight and classifier selection. Comput Mod Life Sci Am Inst Phys Ser 952:258–267
Jensen CA, El-Sharkawi MA, Marks RJ II (2001) Power system security assessment using neural networks: feature selection using fisher discrimination. IEEE Trans Energ Convers 16:757–763
Kramer MA (1991) Nonlinear principal component analysis using autoassociative neural networks. AIChE J 37:233–234
Kramer MA (1992) Autoassociative neural networks. Comput Chem Eng 16:313–328
Kubalík J, Lazanský J (1999) Genetic algorithms and their testing. In: Proceedings of the AIP Conference, New York, USA, pp 217–229
Kwak HG, Kim J (2009) An integrated genetic algorithm complemented with direct search for optimum design of RC frames. Comput Aid Des 41:490–500

Levin RI, Lieven NAJ (1998) Dynamic finite element model updating using simulated annealing and genetic algorithms. Mech Syst Signal Process 12:91–120
Li X, Du G (2012) Inequality constraint handling in genetic algorithms using a boundary simulation method. Comput Oper Res 39:521–540
Lopes V, Park G, Cudney HH, Inman DJ (2000) Impedance-based structural health monitoring with artificial neural networks. J Intell Mater Syst Struct 11:206–216
Lu PJ, Hsu TC (2002) Application of autoassociative neural network on gas-path sensor data validation. J Propul Power 18:879–888
Maia NMM, Silva JMM (1997) Theoretical and experimental modal analysis. Research Studies Press, Letchworth
Marwala T (2000) Fault identification using neural networks and vibration data. PhD thesis, University of Cambridge
Marwala T (2001) Probabilistic fault identification using a committee of neural networks and vibration data. J Aircraft 38:138–146
Marwala T (2002) Finite element updating using wavelet data and a genetic algorithm. AIAA J Aircraft 39:709–711
Marwala T (2003) Fault classification using pseudo modal energies and neural networks. Am Inst Aeronaut Astronaut J 41:82–89
Marwala T (2004) Fault classification using pseudo modal energies and probabilistic neural networks. J Eng Mech 130:1346–1355
Marwala T (2007) Bayesian training of neural network using genetic programming. Pattern Recognit Lett 28:1452–1458
Marwala T (2009) Computational intelligence for missing data imputation, estimation and management: Knowledge optimization techniques. IGI Global Publications, New York
Marwala T (2010) Finite element model updating using computational intelligence techniques. Springer, London
Marwala T, Chakraverty S (2006) Fault classification in structures with incomplete measured data using autoassociative neural networks and genetic algorithms. Curr Sci 90:542–548
Marwala T, Hunt HEM (1999) Fault identification using finite element models and neural networks. Mech Syst Signal Process 13:475–490
Marwala T, de Wilde P, Correia L, Mariano P, Ribeiro R, Abramov V, Szirbik N, Goossenaerts J (2001) Scalability and optimisation of a committee of agents using a genetic algorithm. In: Proceedings of the 2001 international symposium on soft computing and intelligence systems for Industry, Scotland
Michalewicz Z (1996) Genetic algorithms + data structures = evolution programs. Springer, New York
Mitchell M (1996) An introduction to genetic algorithms. MIT Press, Cambridge
Mohamed AK, Nelwamondo FV, Marwala T (2008) Estimation of missing data: neural networks, principal component analysis and genetic algorithms. In: Proceedings of the 12th world multi-conference on systems, cybern and inform, Orlando, Florida, pp 36–41
Møller M (1993) A scaled conjugate gradient algorithm for fast supervised learning. Neural Netw 6:525–533
Mosalman Yazdi HA, Ramli Sulong NH (2001) Optimization of off-centre bracing system using genetic algorithm. J Constr Steel Res 67:1435–1441
Musharavati F, Hamoud ASM (2011) Modified genetic algorithms for manufacturing process planning in multiple parts manufacturing lines. Expert Syst Appl. doi:10.1016/j.eswa.2011.01.129
Neal RM (1993) Probabilistic inference using Markov Chain Monte Carlo methods. Technical Report CRG-TR-93-1, University of Toronto, Toronto, Canada
Nelwamondo FV (2008) Computational intelligence techniques for missing data imputation. PhD thesis, University of the Witwatersrand
Nelwamondo FV, Marwala T (2007) Handling missing data from heteroskedastic and nonstationary data. Lect Notes Comput Sci 449:1297–1306
Oh S, Pedrycz W (2006) Genetic optimization-driven multi-layer hybrid fuzzy neural networks. Simulat Mod Pract Theory 14:597–613

Paluch B, Grédiac M, Faye A (2008) Combining a finite element programme and a genetic algorithm to optimize composite structures with variable thickness. Compos Struct 83:284–294
Park BJ, Choi HR, Kim HS (2003) A hybrid genetic algorithm for the job shop scheduling problems. Comput Ind Eng 45:597–613
Pawar PP, Ganguli R (2003) Genetic fuzzy system for damage detection in beams and helicopter rotor blades. Comput Meth Appl Mech Eng 192:2031
Pendharkar PC, Rodger JA (1999) An empirical study of non-binary genetic algorithm-based neural approaches for classification. In: Proceedings of the 20th international conference on information systems, Charlotte, North Carolina, pp 155–165
Perera R, Ruiz A, Manzano C (2009) Performance assessment of multi-criteria damage identification genetic algorithms. Comput Struct 87:120–127
Reddy RRK, Ganguli R (2003) Structural damage detection in a helicopter rotor using radial basis function neural networks. Smart Struct Mater 12:232–241
Reed RD, Marks RJ II (1999) Neural smithing: supervised learning in feedforward artificial neural networks. MIT Press, Cambridge
Ritter GX, Urcid G, Schmalz MS (2009) Autonomous single-pass end member approximation using lattice auto-associative memories. Neurocomputing 72:2101–2110
Roy T, Chakraborty D (2009) Optimal vibration control of smart fiber reinforced composite shell structures using improved a genetic algorithm. J Sound Vib 319:15–40
Royston TJ, Spohnholtz T, Ellington WA (2000) Use of non-degeneracy in nominally axisymmetric structures for fault detection with application to cylindrical geometries. J Sound Vibr 230:791–808
Sanz J, Perera R, Huerta C (2007) Fault diagnosis of rotating machinery based on auto-associative neural networks and wavelet transforms. J Sound Vibr 302:981–999
Stern H, Chassidim Y, Zofi M (2006) Multi-agent visual area coverage using a new genetic algorithm selection scheme. Eur J Oper Res 175:1890–1907
Suresh S, Omkar SN, Ganguli R, Mani V (2004) Identification of crack location and depth in a centilever beam using a modular neural network approach. Smart Mater Struct 13:907–916
Tettey T, Marwala T (2006) Controlling interstate conflict using neuro-fuzzy modeling and a genetic algorithms. In: Proceedings of the 10th IEEE international conference on intelligent engineering systems, London, UK, pp 30–44
Tu Z, Lu Y (2008) Finite element model updating using artificial boundary conditions with genetic algorithms. Comput Struct 86:714–727
Upadhyaya BR, Eryurek E (1992) Application of neural networks for sensor validation and plant monitoring. Nuclear Technol 97:170–176
Vose MD (1999) The simple genetic algorithm: foundations and theory. MIT Press, Cambridge
Waszczyszyn Z, Ziemianski L (2001) Neural networks in mechanics of structures and materials – new results and prospects of applications. Comput Struct 79:2261–2276
Wu X, Ghaboussi J, Garret JH (1992) Use of neural networks in the detection of structural damage. Comput Struct 42:649–659
Zang C, Imregun M (2001) Combined neural network and reduced FRF techniques for slight damage detection using measured response data. Arch Appl Mech 71:525–536
Zhang H, Ishikawa M (2004) A solution to combinatorial optimization with time-varying parameters by a hybrid genetic algorithm. Int Congr Ser 1269:149–152
Zhou HF, Ni YQ, Ko JM (2011) Structural damage alarming using auto-associative neural network technique: exploration of environment-tolerant capacity and setup of alarming threshold. Mech Syst Signal Process 25:1508–1526

Chapter 10
Condition Monitoring Using Support Vector Machines and Extension Neural Networks Classifiers

10.1 Introduction

The condition monitoring of machines is very important in industry because of the necessity to improve machine reliability and reduce the possible loss of production from machine breakdowns. Condition monitoring is conducted when it is essential to classify the state of a machine and to establish whether it is faulty through observation and analysis (William et al. 1992; Marwala and Vilakazi 2007). *Condition monitoring* is a method of monitoring the operating characteristics of a machine so that the changes and movements of the monitored signals can be used to predict the need for maintenance before a breakdown does happen. Condition monitoring has become progressively more vital in areas such as aerospace engineering where an unpredicted fault can result in a serious accident. Another application of condition monitoring is in manufacturing where manufacturers must identify methods to avoid failures, decrease maintenance costs, minimize downtime, and increase the lifetime of their equipment.

With a reliable condition monitoring procedure, machines can be employed in a more optimal fashion. A maintenance plan follows a schedule to select when maintenance must be performed. This leads to inefficiencies because the maintenance process may be performed unnecessarily early or a failure may occur prior to scheduled maintenance taking place. However, condition monitoring can be applied for condition based maintenance or for predictive maintenance.

Rotating machinery is employed in a number of industrial applications. A common component of a modern rotating machinery is the rolling element bearing. In fact most machine failures are related to bearing failure (Lou and Loparo 2004), which frequently result in protracted downtimes that have economic consequences. As a result, an increasing amount of condition monitoring data are measured and presented to engineers for analysis. However, because of the complexity and mass of data generated relating to all plant items and their health, it is problematic for engineers to handle such data. Because the identification of the vital information from such data is difficult, a reliable and automated diagnostic method permitting

T. Marwala, *Condition Monitoring Using Computational Intelligence Methods: Applications in Mechanical and Electrical Systems*, DOI 10.1007/978-1-4471-2380-4_10,

fairly unskilled operators is necessary to take significant decisions without the need for a condition monitoring specialist. One of the most frequently applied condition monitoring techniques is the vibration-based condition monitoring procedure which is grounded on the principle that all systems produce vibrations. When a machine is operating appropriately, the vibrations are small and constant; however, when faults progress and several dynamic processes change, the vibration spectrum also changes (Marwala 2001).

The robustness of a classification system depends on the usefulness of the extracted features and the reliability and effectiveness of a condition monitoring classification system. This chapter presents three feature selection methods applied for bearing fault diagnosis (Nelwamondo et al. 2006a). These methods are the *Mel-frequency Cepstral Coefficients* (MFCC) technique, which is a time-frequency domain technique; *kurtosis* which is the time-domain procedure and a *fractal dimension analysis* which is also time-domain method that has been of benefit. This chapter also assesses the usefulness of the extracted features for bearing fault diagnosis using the Support Vector Machine (SVM) and the Extension Neural Network (ENN) classifiers. The SVM was selected because it has been applied successfully in many fault complex applications (Msiza et al. 2007; Patel and Marwala 2009; Marivate et al. 2008) and the ENN was selected because of its success in pattern recognition of complex systems (Vilakazi and Marwala 2006; Mohamed et al. 2006).

10.2 Features

The success of a classification system depends on the usefulness of the extracted features that represent a specific machine condition. Previously, substantial research has been conducted into the development of a number of feature extraction methods and condition monitoring systems. Feature extraction methods can be categorized into three domains: the frequency, time-frequency and time-domains (Ericsson et al. 2004). The frequency domain techniques usually include a frequency analysis of the vibration signals and investigate the periodicity of high frequency transients. In this regard, frequency domain approaches explore a train of repetitions arising at any of the frequencies from the faulty regime (Ocak and Loparo 2004). This technique becomes complicated because the periodicity of the signal may be suppressed. These frequency domain methods include the frequency averaging procedure, adaptive noise cancellation and the High Frequency Resonance Technique (HFRT). The HFTR has been used widely for bearing fault detection and diagnosis (Ocak and Loparo 2004). The shortcoming of the HFTR method is that it requires several impact tests to identify the bearing resonance frequency and thus it is computationally expensive (Ocak and Loparo 2004).

McFadden and Smith (1984) presented an envelope analysis, which is a frequency domain method for the detection and diagnosis of bearing faults. The shortcoming of the frequency domain analysis is that it tends to average out transient vibrations. As a result it becomes more sensitive to background noise. To overcome this problem, the time-frequency domain analysis was applied, which expresses how the frequency content of the signal varies with time. Time-frequency domain analysis methods include the Short Time Fourier Transform (STFT), the Wigner-Ville Distribution (WVD) and Wavelet Transform (WT). These techniques are studied in detail in Li et al. (2000).

Time domain approaches normally involve indices that are sensitive to impulsive oscillations, such as peak level, root mean square value, crest factor analysis, Kurtosis analysis, shock pulse counting, time series averaging techniques, and signal enveloping routines (Ocak and Loparo 2004; Li et al. 2000). Ericsson et al. (2004) and Li et al. (2000) demonstrated that the time-domain analysis is less sensitive to suppressions of the periodicity.

A number of feature extraction methods have been applied to vibration-based condition monitoring. Ocak and Loparo (2004); Lou and Loparo (2004), as well as Nikolaou and Antoniadis (2002) have applied wavelet transforms to detect and classify different faults in bearings. Elsewhere, Rojas and Nandi (2006) implemented spectral and statistical features for the classification of bearing faults. Peng et al. (2005) compared the Hilbert-Huang transform with the wavelet transform for a bearing fault diagnosis. Junsheng et al. (2006) presented a feature extraction technique based on empirical mode decomposition process and autoregressive model in roller bearings diagnosis. Baillie and Mathew (1996) applied an autoregressive modeling that does not only classify, but also offers a one-step-ahead prediction of the vibration signal using the previous outputs. H. Yang et al. (2005) implemented a basis pursuit and obtained better results than with wavelet transforms. Altman and Mathew (2001) also applied the discrete wavelet packet analysis to improve the detection and diagnostic effectiveness of rolling element bearing faults. Prabhakar et al. (2002) demonstrated that the Discrete Wavelet Transform (DWT) can be applied to enhance the detection of bearing faults and elsewhere, Antoni and Randall (2006) applied spectral kurtosis to the vibratory surveillance and diagnostics of rotating machines.

The other vital component of a condition monitoring process is a classification system that identifies the operating status of the machine as well as the type of failure. Such a classification system can be categorized into two groups: knowledge-based and data-based models.

Knowledge-based models depend on human-like knowledge of the process and its faults. Knowledge-based models (such as expert systems or decision trees) implement human-like knowledge of the process for fault diagnosis. In fault diagnostics, the human expert could be a person who operates the machine and who has expertise in different categories of faults. The knowledge base can be constructed by conducting interviews with a human operator about fault incidences in the diagnosed machine. Expert systems are normally appropriate for such problems, where the human expert can linguistically explain the solution. Characteristically, human knowledge is imprecise, and how to treat such information has frequently been a problem with traditional expert systems. For instance, the precise limit when the temperature in a sauna is too high is vague in human minds. In fact, it is very difficult to attain sufficiently representative data for the complex and highly non-linear behavior of faulty system to make quantitative models. Knowledge-based models may be applied collectively with simple signal-based diagnostics, if the expert knowledge for the process is available. Nevertheless, it is frequently difficult even for a human expert to differentiate faulty operation from healthy operation. Furthermore, multiple information sources may be needed for reliable decision-making. Therefore, data-based models are the most flexible method for automated condition monitoring.

Data-based models are implemented when the process model is unknown in an analytical form and expert knowledge of the procedure performance under faults is absent. Data-based models can be produced in numerous ways. Neural network based models like the Multilayer Perceptron (MLP) and the Radial Basis Function (RBF) have been applied extensively for bearing condition monitoring. Samanta and Al-Bushi (2003) applied neural networks with time-domain features for the detection of rolling element bearing faults. Elsewhere Yang et al. (2004) implemented the ART-KOHONEN technique for the fault diagnosis of rotating machinery. Kernel-based classifiers such as the Support Vector Machine (SVM) have been applied for bearing fault diagnosis. In this regard, Jack and Nandi (2002) compared support vector machines and neural networks, enhanced by genetic algorithms for fault detection. Samanta (2004) implemented both Artificial Neural Networks (ANN) and SVM with genetic algorithm for bearing fault detection. B.S. Yang et al. (2005) applied multi-class SVM for fault diagnosis of rotating machinery while Rojas and Nandi (2006) applied the SVM for the detection and classification of faults in rolling element bearings. Hu et al. (2007) applied a wavelet transform and SVM ensemble for the fault diagnosis of rotating machines. Furthermore, data-based statistical methods have achieved success in speech recognition and have recently been applied for condition monitoring. Ertunc et al. (2001) applied the Hidden Markov model (HMM) in wear studies of drill bits in a drilling process. Ocak and Loparo (2004), Purushotham et al. (2005), Miao and Makis (2006) and Nelwamondo et al. (2006a) applied the HMM for bearing fault diagnosis.

10.3 Feature Extraction

This section describes a number of features that were extracted from vibration signals of bearing elements: fractal analysis, Mel-frequency cepstral coefficients and kurtosis.

10.3.1 Fractal Dimension

As described in Chap. 2, a *fractal dimension* is a rough geometric shape that can be divided into parts, each of which is nearly a reduced copy of the whole and this characteristic is known as self-similarity. Vibration signals are mostly periodic movements with some level of turbulence. To detect different bearing faults these non-linear, turbulence features must be identified. The non-linear turbulence features of the vibration signal may be quantified using a fractal model (Maragos and Potamianos 1999). The fractal dimension of a compact planar set *F*, also known as the Hausdorff dimension, has its value ranging between 1 and 2. It can be approximated using techniques such as the Minkowski-Bouligand dimension and Box-Counting dimension (Maragos and Potamianos 1999). Fractals have been successfully applied to condition monitoring (Cui and Xu 1999).

Li et al. (2005) applied fractal theory in direct current system grounding fault detection. Elsewhere, Zhao and Guo (2005) applied the fractal dimension technique for detecting a single-phase-to-earth faults. Other applications of the fractal technique were in phase selection (D. Yang et al. 2005), and in aircraft fault detection (Zhang et al. 2001). For this chapter the Box-Counting dimension was used as described in Chap. 2.

10.3.2 Mel-Frequency Cepstral Coefficients (MFCCs)

The *Mel-frequency Cepstral Coefficients* (MFCCs) are coefficients that are derived from a sort of cepstral representation of the signal. They extract both linear and non-linear properties. The frequency bands are evenly spread out on the mel scale and this estimates the human auditory system's response closer than linearly-spaced frequency bands would.

The MFCCs have been applied to fields such as speech recognition and condition monitoring (Marwala and Vilakazi 2007; Nelwamondo et al. 2006b; Mahola et al. 2005; Chen et al. 2011a). The MFCC used features extracted from vibration signals. It can be viewed as a category of wavelet in which frequency scales are put on a linear scale for frequencies less than 1 kHz and on a log scale for frequencies above 1 kHz (Wang et al. 2002). The complex cepstral coefficients attained from this scale are known as the MFCC (Wang et al. 2002). The MFCCs are represented in both time and frequency domains.

Sáenz-Lechón et al. (2011) applied the mel-frequency cepstral analysis for an objective evaluation of perceived roughness and breathiness while Arias-Londoño et al. (2011) applied the mel-frequency cepstral coefficients for the automatic detection of pathological voices. Other applications of the mel-frequency cepstral coefficients were for the classification of spoken letters (Rozali et al. 2011), and in the low bit-rate coding of speech (Boucheron et al. 2011).

10.3.3 Kurtosis

Kurtosis is the quantification of the "peakedness" in the probability distribution of a real-valued, random variable. In this chapter, kurtosis is intended to deal with the occasional spiking of vibration data caused by some types of faults. Kurtosis features of vibration data have been applied in the monitoring of tool condition by El-Wardany et al. (1996). Elsewhere, Cai et al. (2011) successfully applied kurtosis in detecting roller bearing faults. Immovilli et al. (2009) successfully applied kurtosis for the detection of generalized-roughness bearing faults. Elsewhere, Tao et al. (2008) applied Kurtosis for the detection of faults for one class of bearings. Finally, kurtosis was applied for fault detection and diagnosis in rolling element bearings by Sawalhi et al. (2007).

The success of using kurtosis in vibration signals is a consequence of the fact that the vibration signals of a system under stress or having defects are different from those of a normal system. The spiking of the vibration signal changes when there are faults in the system. Kurtosis is a quantification of the sharpness of the peak and is defined as the normalized fourth-order central moment of the signal (Wang et al. 2001). The kurtosis value is beneficial in identifying transients and spontaneous events within vibration signals (Wang 2001) and is an accepted criterion for fault detection. The kurtosis value is computed as the normalized square of the second moment. A high value of kurtosis indicates a sharp distribution peak and shows that the signal is impulsive in nature (Altman and Mathew 2001).

10.4 Classification Techniques

For this chapter two classification techniques were applied, together with the features described in the previous section for the detection of faults in bearings. These classification techniques are described in this section *viz.*, the support vector machines and extension neural networks.

10.4.1 Support Vector Machines (SVMs)

Support vector machines are supervised learning approaches applied mostly for classification. They are derived from the theory of statistical learning and were first proposed by Vapnik (1995). Shen et al. (2005) applied a SVM-based color image watermarking method that operated by applying the information supplied from the reference positions and the watermark, which was adaptively embedded into the blue channel of the host image, taking the human visual system into account. Other implementations of SVMs to model complicated systems include Marwala et al. (2006) who applied SVMs in the fault classification of mechanical systems, Msiza et al. (2007) who used SVMs in forecasting a water-demand time-series and Marwala and Lagazio (2011) who applied SVMs in the modeling of militarized interstate conflict.

Chen et al. (2011b) used SVMs to approximate monthly solar radiation and their results demonstrated that SVMs perform better than traditional approaches (such as neural networks) in predicting solar radiation. Yeh et al. (2011) applied SVMs in the recognition of counterfeit banknotes. Each banknote was separated into segments and the luminance histograms of the segments were used as the inputs to the SVM model with each segment was paired with its own kernels. When the technique was tested on Taiwanese banknotes, they showed that their technique performed better than methods such as the single-kernel SVM.

Tellaeche et al. (2009) used support vector machines and computer vision for weed identification. Elsewhere, Lin et al. (2011) applied SVMs to predict business failures based on previous financial data. Their results showed that their method gave a good classification rate. Li-Xia et al. (2011) implemented SVMs and particle swarm optimization for tax forecasting and their results showed that the SVM model performs well.

The application of SVMs has also been extended to regression analysis problems, thus resulting in the term Support Vector Regression (SVR) (Gunn 1997; Chuang 2008). Pires and Marwala (2004) applied SVMs for option pricing and extended these with a Bayesian framework. Elsewhere, Gidudu et al. (2007) used SVMs in image classification.

Thissen et al. (2004) applied SVMs for spectral regression cases, and Üstün et al. (2007) visualized and interpreted SVM models. Other applications of SVMs include the prediction of jet penetration depth (Wang et al. 2010), tool wear identification (Tao and Tao 2010), ozone concentration (Ortiz-García et al. 2010), the identification of people (Palanivel and Yegnanarayana 2008), chemical compound analysis (Zhou et al. 2006), response modeling (Kim et al. 2008), and the real-time prediction of order flow times (Alenczi et al. 2007).

For SVMs, a data point is conceptualized as a p-dimensional vector. The objective is to separate such points with a p-1-dimensional hyperplane, known as a *linear classifier*. There are many hyperplanes that can be created. Some of these include the one that exhibits the largest separation, also called the *margin*, between the two classes. The selected hyperplane can be chosen so that the distance from it to the nearest data point on both sides is maximized. This is then known as the *maximum-margin hyperplane*. The classification problem can then be stated as estimating a function $f : R^N \rightarrow \{-1, 1\}$ dependent on input-output training data, where an independently distributed, unknown probability distribution $P(\boldsymbol{x}, y)$ is chosen such that f can classify unseen $(\boldsymbol{x}, y)$ data (Müller et al. 2001; Habtemariam 2006; Marwala and Lagazio 2011). The ideal function minimizes the expected error (risk) and can be expressed mathematically as follows (Vapnik 1995; Habtemariam 2006; Habtemariam et al. 2005; Marwala and Lagazio 2011):

$$R[f] = \int l(f(\mathbf{x}), y) dP(\mathbf{x}, y) \tag{10.1}$$

Here, l indicates a loss function (Müller et al. 2001). Since the fundamental probability distribution P is unknown, Eq. 10.1 cannot be solved implicitly.

The best route is to identify an upper bound for the risk function which is given mathematically as follows (Vapnik 1995; Müller et al. 2001; Marwala and Lagazio 2011):

$$R[f] = R[f]_{emp} + \sqrt{\frac{h\left(\ln\frac{2n}{h} + 1\right) - \ln\left(\frac{\delta}{4}\right)}{n}} \tag{10.2}$$

Here $h \in \mathrm{N}^+$ is the Vapnik-Chervonenkis (*VC*) dimension of $f \in F$ and $\delta > 0$. The *VC* dimension of a function class F is defined as the biggest number of h coordinates that can be divided in all possible ways by means of functions of that class (Vapnik 1995). The empirical error $R[f]_{emp}$ is a training error given by (Vapnik 1995; Habtemariam 2006; Marwala and Lagazio 2011):

$$R[f]_{emp} = \frac{1}{n}\sum_{i+1}^{n} l(f(x_i), y_i) \tag{10.3}$$

This assumes that the training sample is linearly separable by a hyperplane of the form (Vapnik 1995; Habtemariam 2006; Marwala and Lagazio 2011):

$$f(\mathbf{x}) = \langle \mathbf{w}, \mathbf{x} \rangle + b \;\; with \;\; \mathbf{w} \in \chi, b \in \Re \tag{10.4}$$

Here $\langle .,. \rangle$ denotes the dot product, $\boldsymbol{w}$ is an adjustable weight vector and b is an offset (Müller et al. 2001). The objective of the learning process, as propounded by Vapnik and Lerner (1963) is to identify the hyperplane with the maximum margin of separation from the class of dividing hyperplanes. Nonetheless, since practical data usually exhibit complex properties which cannot be divided linearly, more complex classifiers are essential. To circumvent the complexity of the nonlinear classifiers, the concept of linear classifiers in a feature space can be introduced. SVMs attempt to identify a linear separating hyperplane by initially mapping the input space into a higher dimensional feature space F. This suggests that each training example $\boldsymbol{x}_i$ be substituted with $(\boldsymbol{x}_i)$ to give (Vapnik 1995; Habtemariam 2006):

$$Y_i\left(\mathbf{w}.\Phi(\mathbf{x}_i) + b\right), i = 1, 2, ..., n \tag{10.5}$$

The *VC* dimension,*h*, in the feature space F is constrained subject to $h \leq ||W||^2 R^2 + 1$ where R is the radius of the smallest sphere around the training data (Müller et al. 2001; Habtemariam 2006). Consequently, minimizing the expected risk can be expressed as an optimization problem as follows (Burges 1998; Müller et al. 2001; Schölkopf and Smola 2003; Marwala and Lagazio 2011):

$$\text{Minimize } (W, b)\; \frac{1}{2}||\mathbf{w}||^2 \tag{10.6}$$

subject to:

$$c_i(\mathbf{w}, \mathbf{x}_i - b) \geq 1, i = 1, ..., n \tag{10.7}$$

Equations 10.6 and 10.7 are jointly called the *quadratic programming problem* because it is the problem of optimizing a quadratic function of a number of variables subject to linear constraints on these variables (Schölkopf and Smola 2003). From the expressions:

$$\|\mathbf{w}\|^2 = \mathbf{w}.\mathbf{w} \tag{10.8}$$

$$\mathbf{w} = \sum_{i=0}^{n} \alpha_i c_i \mathbf{x}_i \tag{10.9}$$

it can be shown that the dual nature of the support vector machines can, by maximizing of α_i, be written in Lagrangian form as follows (Schölkopf and Smola 2003):

$$\begin{aligned} L(\alpha) &= \sum_{i=1}^{n} \alpha_i - \frac{1}{2} \sum_{i,j} \alpha_i \alpha_j c_i c_j \mathbf{x}_i^T \mathbf{x}_j \\ &= \sum_{i=1}^{n} \alpha_i - \frac{1}{2} \sum_{i,j} \alpha_i \alpha_j c_i c_j k(\mathbf{x}_i, \mathbf{x}_j), i = 1, ..., n \end{aligned} \tag{10.10}$$

Subject to:

$$\alpha_i \geq 0, i = 1, ..., n \tag{10.11}$$

and to the constraint from the minimization in b:

$$\alpha_i \geq 0, i = 1, ..., n \tag{10.12}$$

and subject to the following constraints:

$$\sum_{i=1}^{n} \alpha_i c_i = 0 \tag{10.13}$$

Here the kernel is (Müller et al. 2001):

$$k(\mathbf{x}_i, \mathbf{x}_j) = \mathbf{x}_i \cdot \mathbf{x}_j \tag{10.14}$$

10.4.1.1 Soft Margin

Cortes and Vapnik (1995) presented an improved maximum margin idea that incorporated mislabeled data points. If there is no hyperplane that can exactly divide the "yes" and "no" data points, the *Soft Margin* technique will select a hyperplane that divides the data points as efficiently as possible, still maximizing the distance to the nearest, neatly divided data points. The method incorporates slack variables, γ_i which quantify the degree of misclassification of the data point as follows (Cortes and Vapnik 1995):

$$c_i\left(\mathbf{w}\cdot\mathbf{x}_i - b\right) \geq 1 - \gamma_i, 1 \leq i \leq n \tag{10.15}$$

A function which penalizes non-zero γ_i augments the objective and, consequently, the optimization exhibits a compromise between a large margin and a small error penalty. If a linear penalty function is assumed, the optimization problem can then be written by minimizing $\boldsymbol{w}$ and γ_i through the following function (Cortes and Vapnik 1995):

$$\frac{1}{2}\|\mathbf{w}\|^2 + C\sum_{i=1}^{n}\gamma_i \tag{10.16}$$

subject to:

$$c_i\left(\mathbf{w}\cdot\mathbf{x}_i - b\right) \geq 1 - \gamma_i, \gamma_i \geq 0, i = 1, ..., n \tag{10.17}$$

In Eq. 10.16, C is the capacity. Equations 10.16 and 10.17 can be expressed in a Lagrangian form by optimizing the following equation in terms of $\boldsymbol{w}$, γ, b, α and β (Cortes and Vapnik 1995):

$$\min_{\{w\},\gamma,b}\max_{\alpha,\beta}$$

$$\left\{\frac{1}{2}\|\mathbf{w}\|^2 + C\sum_{i=1}^{n}\gamma_i - \sum_{i=1}^{n}\alpha_i\left[c_i\left(\mathbf{w}\cdot\mathbf{x}_i - b\right) - 1 + \gamma_i\right] - \sum_{i=1}^{n}\beta_i\gamma_i\right\} \tag{10.18}$$

Here $\alpha_i, \beta_i \geq 0$. The advantage of a linear penalty function is that the slack variables are removed from the dual problem. As a result, C only appears as a supplementary constraint on the Lagrange multipliers. The application of non-linear penalty functions to reduce the impact of outliers on the classifier has been applied in the past, but it makes the optimization problem non-convex and it is difficult to identify a global solution.

10.4.1.2 Non-Linear Classification

To apply the linear SVM technique for producing non-linear classifiers, the kernel trick was applied (Aizerman et al. 1964) to the maximum-margin hyperplanes (Boser et al. 1992). In this method the dot product is substituted with a non-linear kernel function to fit the maximum-margin hyperplane in a transformed feature space. Although this dot product transformation may be non-linear, the transformed space may be of high dimensions. For instance, when a Gaussian radial basis function kernel is applied, the resultant feature space is a Hilbert space of infinite dimension. Some useful kernel functions include (Vapnik 1995; Müller et al. 2001; Marwala and Lagazio 2011):

The Radial Basis Function,

$$k\left(\mathbf{x}_i, \mathbf{x}_j\right) = \exp\left(-\gamma \left\|\mathbf{x}_i - \mathbf{x}_j\right\|^2\right), \gamma > 0 \tag{10.19}$$

The Polynomial (homogeneous),

$$k\left(\mathbf{x}_i, \mathbf{x}_j\right) = \left(\mathbf{x}_i \cdot \mathbf{x}_j\right)^d \tag{10.20}$$

The Polynomial (inhomogeneous),

$$k\left(\mathbf{x}_i, \mathbf{x}_j\right) = \left(\mathbf{x}_i \cdot \mathbf{x}_j + 1\right)^d \tag{10.21}$$

The Hyperbolic tangent,

$$k\left(\mathbf{x}_i, \mathbf{x}_j\right) = \tanh\left(\varepsilon \mathbf{x}_i \cdot \mathbf{x}_j + b\right), \varepsilon > 0; b < 0 \tag{10.22}$$

The variables of the maximum-margin hyperplane can be identified by optimizing the objective equation through applying an interior point technique that identifies a solution for the Karush-Kuhn-Tucker (KKT) conditions of the primal and dual problems (Kuhn and Tucker 1951; Karush 1939). To avoid solving a linear system, including the large kernel matrix, a low rank approximation to the matrix can be applied to apply the kernel trick. The Karush–Kuhn–Tucker conditions are necessary to optimize a non-linear programming problem, for the satisfaction of a particular regularity condition. For the given problem listed below:

$$Minimize : f(\mathbf{x}) \tag{10.23}$$

subject to:

$$g_i\left(\mathbf{x}\right) \leq 0; h_j\left(\mathbf{x}\right) = 0 \tag{10.24}$$

Here, g_i is the ith inequality constraint and h_i is the ith equality constraint. The Karush–Kuhn–Tucker method allows the inequality constraints by generalizing the technique of Lagrange multipliers which allow only equality constraints. The necessary conditions for the KKT are (Kuhn and Tucker 1951; Karush 1939; Marwala and Lagazio 2011):

Stationary,

$$\nabla f\left(\mathbf{x}^{*}\right)+\sum_{i=1}^{m}\mu_{i}\nabla g_{i}\left(\mathbf{x}^{*}\right)+\sum_{j=1}^{l}\lambda_{j}\nabla h_{j}\left(\mathbf{x}^{*}\right)=0, i=1,...,m; j=1,...,l \tag{10.25}$$

Primal and dual feasibility as well as complementary slackness,

$$\begin{aligned} g_{i}\left(\mathbf{x}^{*}\right) &\leq 0, i=1,...,m \\ h_{j}\left(\mathbf{x}^{*}\right) &= 0; j=1,...,l \\ \mu_{i} &\geq 0, i=1,...,m \\ \mu_{i}g_{i}\left(\mathbf{x}^{*}\right) &= 0, i=1,...,m \end{aligned} \tag{10.26}$$

The KKT method can be viewed as a generalized form of the Lagrangian method obtained by setting $m=0$. In some cases, the necessary conditions are also sufficient for optimization. Nevertheless, in many circumstances the necessary conditions are not sufficient for optimization and additional information, for instance the second derivative, is necessary. The necessary conditions are sufficient for optimization if the cost function f and the inequality constraints g_j are continuously differentiable, convex functions and the equality constraints g_j are functions which have constant gradients.

10.4.2 Extension Neural Networks

An Extension Neural Network (ENN) is a classification system that is based on concepts from neural networks and extension theory as shown in Fig. 10.1 (Wang and Hung 2003; Mohamed et al. 2006; Vilakazi and Marwala 2006). Lu (2010) successfully applied ENN for fault diagnosis, while Zhang et al. (2010) applied extension neural networks for the condition monitoring of the equipment. Wang et al. (2009) applied extension neural networks for the classification of brain MRI while Chao et al. (2009) applied ENN for tracking the maximum power point. Other applications of ENN were in infringement lawsuits (Lai and Che 2009), in intelligent traffic light control (Chao et al. 2008) and in the condition monitoring of transformer bushings (Miya et al. 2008).

The extension theory implements a distance measurement for classification processes, and the neural network entrenches the relevant features of learning capability. The classifier is ideal for classification problems where there are patterns with an extensive range of continuous inputs and a discrete output indicating which

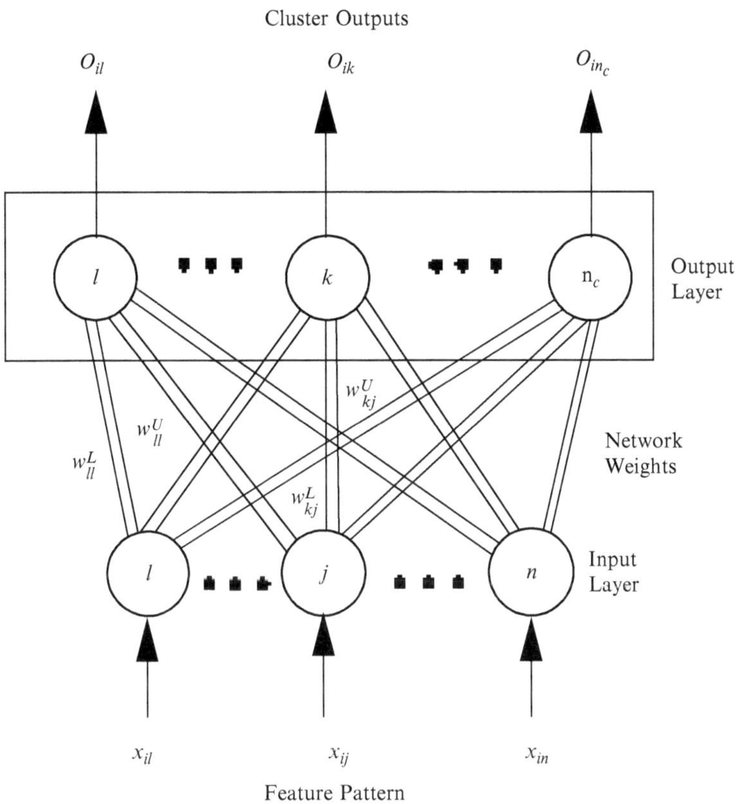

Fig. 10.1 Extension neural network

class the pattern is an element of. ENN encompasses an input layer and an output layer. The input layer nodes accept an input feature pattern and use a set of weighted parameters to produce an image of the input pattern. There are two connection weights between input nodes and output nodes: one connection represents the *lower bound* for this classical domain of features and the other represents the *upper bound.* The complete network is therefore represented by a matrix of weights for the upper and lower limits of the features for each class. These are W_U and W_L, respectively. A third matrix representing the cluster centers is also defined as (Wang and Hung 2003):

$$z = \frac{W_u + W_l}{2} \tag{10.27}$$

The ENN applies supervised learning, which tunes the weights of the ENN to attain a good clustering performance by minimizing the clustering error. The network is trained by adjusting the network weights and recalculating the network centers for each training pattern, subject to reducing the extension distance (ED) of that pattern to its labelled cluster. Each training pattern adjusts the network weights and the centers by quantities that depend on the learning rate. Generally, the weight update for a variable x_i is (Wang and Hung 2003):

$$w^{new} = w^{old} - \eta(x_i - w^{old}) \tag{10.28}$$

Here, η is the learning rate and w can either be the upper or the lower weight matrices of the network centers. It can be shown that for t training patterns for a particular class C, the weight is given by (Mohamed et al. 2006):

$$w^c(t) = (1 - \eta)w^c(0) - \eta \sum (1 - \eta)^{t-1} {x_i}^c \tag{10.29}$$

This equation reveals how each training pattern reinforces the learning in the network by having the most recent signal govern only a fraction of the current value. This equation demonstrates that there is no convergence of the weight values, as the learning process is adaptive and reinforcing and indicates the significance of the learning rate, η. Small values of η necessitate many training iterations, while high values may cause an oscillatory behavior of the network weights, causing poor classification performance.

10.5 Example Vibration Data

The work for this chapter was based on the data obtained from the Case Western Reserve University website (Loparo 1998). The set-up for the corresponding experiment was made up of a Reliance Electric 2HP IQPreAlert connected to a dynamometer. Faults of size 0.007, 0.014, 0.021 and 0.028 in. were introduced into the drive-end bearing of a motor using the Electric Discharge Machining (EDM) technique. These faults were introduced separately at the inner raceway, rolling element and outer raceway. An impulsive force was applied to the motor shaft and the resulting vibration was measured using two accelerometers, one mounted on the motor housing and the other on the outer race of the drive-end bearing. All signals were recorded at a sampling frequency of 12 kHz.

10.6 Application to Bearing Condition Monitoring

Figure 10.2 shows samples of bearing vibration signals for the four bearing conditions (Marwala and Vilakazi 2007). Features were then extracted to classify faults.

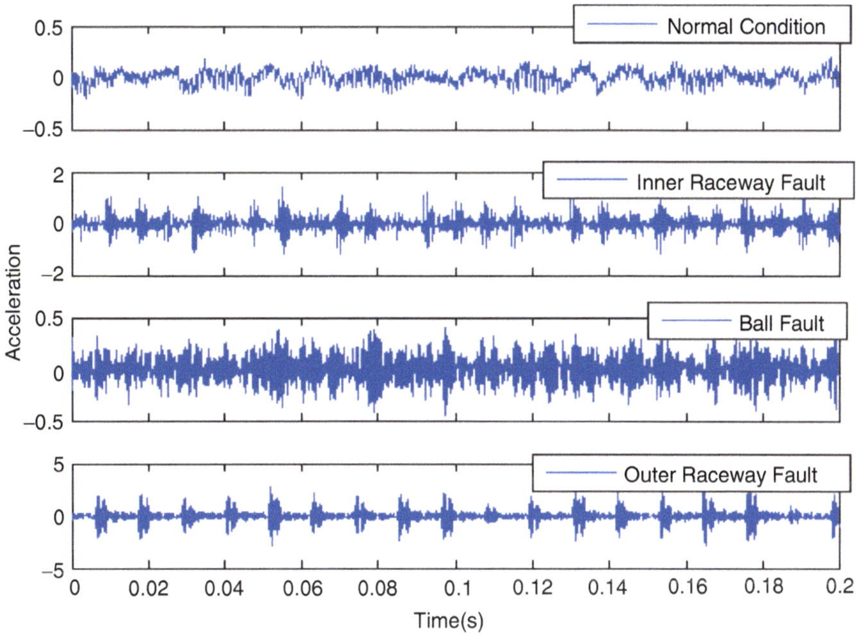

Fig. 10.2 Vibration signals on the bearing under: normal conditions, with inner race fault, outer race fault and ball fault

The optimal classifier parameters were identified using trial and error. The optimum SVM architecture applied a polynomial kernel function with a degree of 5. The ENN architecture with an optimal learning rate of 0.219 was used.

The first set of investigations assessed the effectiveness of the time-domain fractal dimension based feature-extraction using vibration signal condition monitoring. Figure 10.3 shows the Multi-scale Fractal Dimension (MFD) feature vector which shows the bearing's fault-specific information. Figure 10.3 shows that the presented feature extraction method does indeed extract fault defining features which can be applied to classify the different bearing conditions. Nonetheless, the optimum size of the MFD must be found.

Figure 10.4 shows the change of the system accuracy with a change to the MFD size. This figure shows that the size of MFD does not affect the classification accuracy of SVM and ENN.

Using the optimum SVM and ENN architecture together with MFD, the confusion matrix that was obtained for different bearing faults is presented for the SVM and ENN classifiers in Table 10.1.

In further investigating the use of MFCC with SVM and ENN, it was observed that varying the number of MFCCs has no impact on the classification rate of the

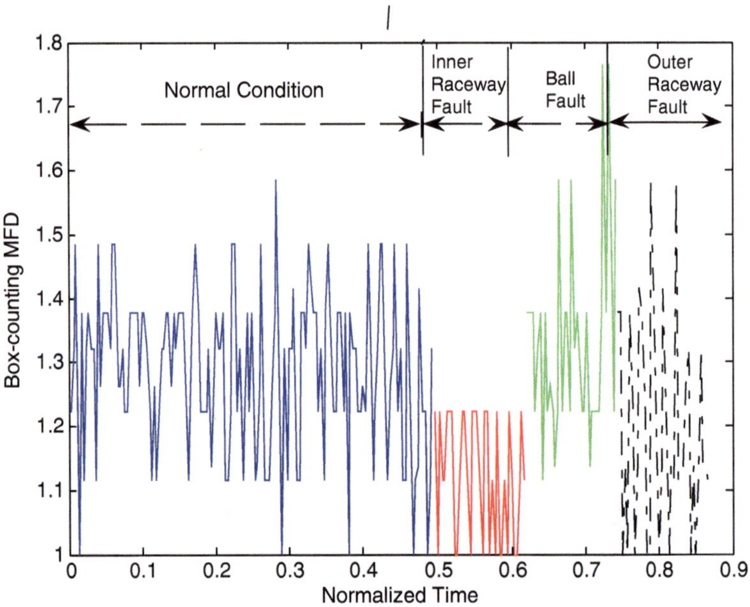

Fig. 10.3 MFD feature extraction comparison for the normal, inner, outer and ball fault for the 1 s vibration signal

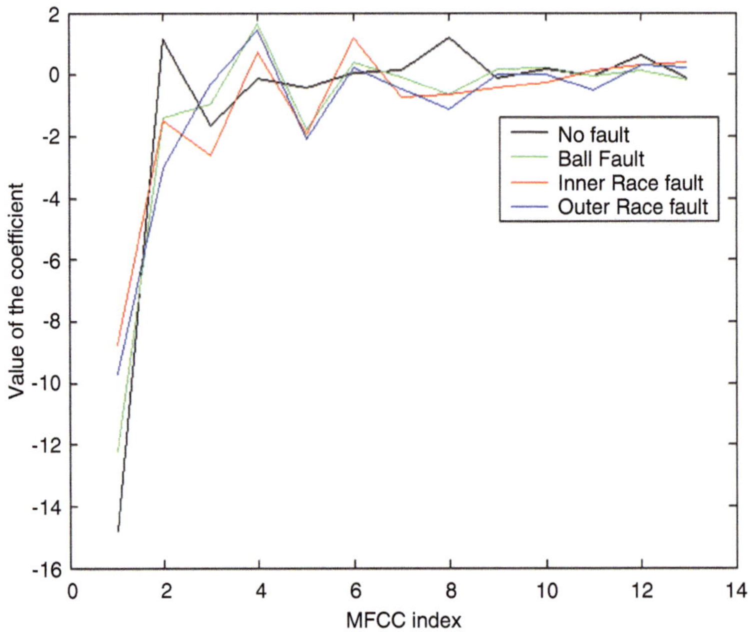

Fig. 10.4 MFCC values corresponding to different fault conditions

Table 10.1 The confusion matrix for the SVM, HMM, GMM, and ENN classifiers used with fractal features

	SVM				ENN			
	Normal	Inner	Outer	Ball	Normal	Inner	Outer	Ball
Normal	100	0	0	0	100	0	0	0
Inner	0	100	0	0	0	100	0	0
Outer	0	0	100	0	0	0	100	0
Ball	0	0	0	100	0	0	0	100

Table 10.2 The confusion matrix for the SVM and ENN classifiers used with MFCC features

	SVM				ENN			
	Normal	Inner	Outer	Ball	Normal	Inner	Outer	Ball
Normal	100	0	0	0	100	0	0	0
Inner	0	100	0	0	0	100	0	0
Outer	0	0	100	0	0	0	100	0
Ball	0	0	0	100	0	0	0	100

Table 10.3 Summary of classification results

	SVM (%)	ENN (%)
Fractal features	100	100
MFCC	100	100
MFCC + kurtosis	100	100

classifiers, as shown Fig. 10.4. It was also observed that 13 MFCCs gave optimal results and that increasing the number of MFCC above 13 did not improve the classification results Table 10.2.

The overall classification results are summarized in Table 10.3.

10.7 Conclusion

The chapter investigated two vital requirements for an automated condition monitoring system. The first requirement was for a feature-extraction procedure that could successfully extract the condition specific features. The second requirement was for a classification system that could effectively classify the machine conditions. This chapter gave a review of three feature extraction methods that are used for condition monitoring. These methods are fractal analysis, mel-frequency cepstral coefficients and kurtosis. The effectiveness of the extracted features were tested using two classifiers. These were support vector machines and the extension neural network. The presented system gave very good results for the fault diagnosis of bearings.

References

Aizerman M, Braverman E, Rozonoer L (1964) Theoretical foundations of the potential function method in pattern recognition learning. Autom Remote Control 25:821–837

Alenezi A, Moses SA, Trafalis TB (2007) Real-time prediction of order flowtimes using support vector regression. Comp Oper Res 35:3489–3503

Altman J, Mathew J (2001) Multiple band-pass autoregressive demodulation for rolling element bearing fault diagnosis. Mech Syst Signal Process 15:963–997

Antoni J, Randall RB (2006) The spectral kurtosis: application to the vibratory surveillance and diagnostics of rotating machines. Mech Syst Signal Process 20:308–331

Arias-Londoño JD, Godino-Llorente JI, Markaki M, Stylianou Y (2011) On combining information from modulation spectra and mel-frequency cepstral coefficients for automatic detection of pathological voices. Logoped Phoniatr Vocol 36:60–69

Baillie DC, Mathew J (1996) A comparison of autoregressive modeling techniques for fault diagnosis of rolling element bearings. Mech Syst Signal Process 10:1–17

Boser BE, Guyon IM, Vapnik VN (1992) A training algorithm for optimal margin classifiers. In: Haussler D (ed) 5th annual ACM workshop on COLT. ACM Press, Pittsburgh

Boucheron LE, Leon PLD, Sandoval S (2011) Hybrid scalar/vector quantization of mel-frequency cepstral coefficients for low bit-rate coding of speech. In: Proceedings of the data compression conference, Snowbird, Utah, pp 103–112

Burges C (1998) A tutorial on support vector machines for pattern recognition. Data Min Knowl Disc 2:121–167

Cai YP, Li AH, Shi LS, Bai XF, Shen JW (2011) Roller bearing fault detection using improved envelope spectrum analysis based on EMD and spectrum kurtosis. J Vibr Shock 30:167–172+191

Chao KH, Lee RH, Wang MH (2008) An intelligent traffic light control based on extension neural network. Lect Notes Comput Sci 5177:17–24

Chao KH, Li CJ, Wang MH (2009) A maximum power point tracking method based on extension neural network for PV systems. Lect Notes Comput Sci 5551:745–755

Chen JL, Liu HB, Wu W, Xie DT (2011a) Estimation of monthly solar radiation from measured temperatures using support vector machines – a case study. Renew Energ 36:413–420

Chen N, Xiao HD, Wan W (2011b) Audio hash function based on non-negative matrix factorisation of mel-frequency cepstral coefficients. IET Info Sec 5:19–25

Chuang CC (2008) Extended support vector interval regression networks for interval input–output data. Info Sci 178:871–891

Cortes C, Vapnik V (1995) Support-vector networks. Mach Learn 20:273–297

Cui R, Xu D (1999) Detection of minor faults using both fractal and artificial neural network techniques. J China Univ Min Technol 28:258–265

El-Wardany TI, Gao D, Elbestawi MA (1996) Tool condition monitoring in drilling using vibration signature analysis. Int J Mach Tool Manufact 36:687–711

Ericsson S, Grip N, Johansson E, Persson LE, Sjöberg R, Strömberg JO (2004) Towards automatic detection of local bearing defects in rotating machines. Mech Syst Signal Process 19:509–535

Ertunc HM, Loparo KA, Ocak H (2001) Tool wear condition monitoring in drilling operations using hidden Markov models. Int J Mach Tool Manufact 41:1363–1384

Gidudu A, Hulley G, Marwala T (2007) Image classification using SVMs: one-against-one vs One-against-all. In: Proceedings of the 28th Asian conference on remote sensing, CD-Rom, Kuala Lumpur, Malaysia

Gunn SR (1997) Support vector machines for classification and regression. ISIS technical report, University of Southampton

Habtemariam E (2006) Artificial intelligence for conflict management. MSc thesis, University of the Witwatersrand

Habtemariam E, Marwala T, Lagazio M (2005) Artificial intelligence for conflict management. In: Proceedings of the IEEE international joint conference on neural networks, Montreal, Canada, pp 2583–2588

Hu Q, He Z, Zhang Z, Zi Y (2007) Fault diagnosis of rotating machine based on improved wavelet package transform and SVMs ensemble. Mech Syst Signal Process 21:688–705
Immovilli F, Cocconcelli M, Bellini A, Rubini R (2009) Detection of generalized-roughness bearing fault by spectral-kurtosis energy of vibration or current signals. IEEE Trans Ind Electron 56:4710–4717
Jack LB, Nandi AK (2002) Fault detection using support vector machines and artificial neural networks, augmented by genetic algorithms. Mech Syst Signal Process 16:373–390
Junsheng C, Dejie Y, Yu Y (2006) A fault diagnosis approach for roller bearings based on EMD method and AR model. Mech Syst Signal Process 20:350–362
Karush W (1939) Minima of functions of several variables with inequalities as side constraints. MSc thesis, University of Chicago
Kim D, Lee H, Cho S (2008) Response modeling with support vector regression. Expert Syst Appl 34:1102–1108
Kuhn HW, Tucker AW (1951) Nonlinear programming. In: Proceedings of the 2nd Berkeley symposium, Berkeley, pp 481–492
Lai YH, Che HC (2009) Integrated evaluator extracted from infringement lawsuits using extension neural network accommodated to patent assessment. Int J Comp Appl Technol 35:84–96
Li B, Chow MY, Tipsuwan Y, Hung JC (2000) Neural-network-based motor rolling bearing fault diagnosis. IEEE Trans Ind Electron 47:1060–1068
Li DH, Wang JF, Shi LT (2005) Application of fractal theory in DC system grounding fault detection. Autom Electric Power Syst 29:53–56+84
Lin F, Yeh CC, Lee MY (2011) The use of hybrid manifold learning and support vector machines in the prediction of business failure. Knowledge-Based Syst 24:95–101
Li-Xia L, Yi-Qi Z, Liu XY (2011) Tax forecasting theory and model based on SVM optimized by PSO. Expert Syst Appl 38:116–120
Loparo KA (1998) Bearing data centre. Case Western Reserve University. http://www.eecs.cwru.edu/laboratory/bearing. Accessed 19 Nov 2005
Lou X, Loparo KA (2004) Bearing fault diagnosis based on wavelet transform and fuzzy inference. Mech Syst Signal Process 18:1077–1095
Lu M (2010) The study of fault diagnosis algorithm based on extension neural network. In: Proceedings of the 2nd IEEE international conference on information and financial engineering, Chongqing, China, pp 447–450
Mahola U, Nelwamondo FV, Marwala T (2005) HMM sub-band based speaker identification. In: Proceedings of the 16th annual symposium of the Pattern Recognition Society of South Africa, Langebaan, pp 123–128
Maragos P, Potamianos A (1999) Fractal dimensions of speech sounds: computation and application to automatic speech recognition. J Acoust Soc Am 105:1925–1932
Marivate VN, Nelwamondo VF, Marwala T (2008) Investigation into the use of autoencoder neural networks, principal component analysis and support vector regression in estimating missing HIV data. In: Proceedings of the 17th world congress of the international federation of automatic control, Seoul, Korea, pp 682–689
Marwala T (2001) Fault identification using neural network and vibration data. PhD thesis, University of Cambridge
Marwala T, Lagazio M (2011) Militarized conflict modeling using computational intelligence techniques. Springer, London
Marwala T, Vilakazi CB (2007) Condition monitoring using computational intelligence. In: Laha D, Mandal P (eds) Handbook on computational intelligence in manufacturing and production management. IGI Publishers, Hershey
Marwala T, Chakraverty S, Mahola U (2006) Fault classification using multi-layer perceptrons and support vector machines. Int J Eng Simul 7:29–35
McFadden PD, Smith JD (1984) Vibration monitoring of rolling element bearings by high frequency resonance technique – a review. Tribol Int 77:3–10
Miao Q, Makis V (2006) Condition monitoring and classification of rotating machinery using wavelets and hidden Markov models. Mech Syst Signal Process 21:840–855

Miya WS, Mpanza LJ, Marwala T, Nelwamondo FV (2008) Condition monitoring of oil-impregnated paper bushings using extension neural network, Gaussian mixture and hidden Markov models. In: Proceedings of the IEEE international conference on systems, man, and cybernetics, Tucson, pp 1954–1959

Mohamed S, Tettey T, Marwala T (2006) An extension neural network and genetic algorithm for bearing fault classification. In: Proceedings of the IEEE international joint conference on neural networks, Vancouver, Canada, pp 7673–7679

Msiza IS, Nelwamondo FV, Marwala T (2007) Artificial neural networks and support vector machines for water demand time series forecasting. In: Proceedings of the IEEE international conference on systems, man, and cybernetics, Montreal, Canada, pp 638–643

Müller KR, Mika S, Ratsch G, Tsuda K, Scholkopf B (2001) An introduction to kernel-based learning algorithms. IEEE Trans Neural Netw 12:181–201

Nelwamondo FV, Mahola U, Marwala T (2006a) Multi-scale fractal dimension for speaker identification system. WSEAS Trans Syst 5:1152–1157

Nelwamondo FV, Marwala T, Mahola U (2006b) Early classifications of bearing faults using hidden Markov models, Gaussian mixture models, mel-frequency cepstral coefficients and fractals. Int J Innov Comput Info Control 2:281–1299

Nikolaou NG, Antoniadis LA (2002) Rolling element bearing fault diagnosis using wavelet packets. NDT&E Intl 35:197–205

Ocak H, Loparo KA (2004) Estimation of the running speed and bearing defect frequencies of an induction motor from vibration data. Mech Syst Signal Process 18:515–533

Ortiz-García EG, Salcedo-Sanz S, Pérez-Bellido ÁM, Portilla-Figueras JA, Prieto L (2010) Prediction of hourly O_3 concentrations using support vector regression algorithms. Atmos Environ 44:4481–4488

Palanivel S, Yegnanarayana B (2008) Multimodal person authentication using speech, face and visual speech [modalities]. Comput Vis Image Underst 109:44–55

Patel PB, Marwala T (2009) Genetic algorithms, neural networks, fuzzy inference system, support vector machines for call performance classification. In: Proceedings of the IEEE international conference on machine learning and applications, Miami, Florida, pp 415–420

Peng ZK, Tse PW, Chu FL (2005) A comparison study of improved Hilbert–Huang transform and wavelet transform: application to fault diagnosis for rolling bearing. Mech Syst Signal Process 19:974–988

Pires M, Marwala T (2004) Option pricing using neural networks and support vector machines. In: Proceedings of the IEEE international conference on systems, man, and cybernetics, The Hague, Netherlands, pp 1279–1285

Prabhakar S, Mohanty AR, Sekhar AS (2002) Application of discrete wavelet transform for detection of ball bearing race faults. Tribol Int 35:793–800

Purushotham V, Narayanana S, Prasadb SAN (2005) Multi-fault diagnosis of rolling bearing elements using wavelet analysis and hidden Markov model based fault recognition. NDT&E Int 38:654–664

Rojas A, Nandi AK (2006) Practical scheme for fast detection and classification of rolling-element bearing faults using support vector machines. Mech Syst Signal Process 20:1523–1536

Rozali MF, Yassin IM, Zabidi A, Mansor W, Tahir NMD (2011) Application of Orthogonal Least Square (OLS) for selection of Mel frequency cepstrum coefficients for classification of spoken letters using MLP classifier. In: Proceedings of the IEEE 7th international colloquium on signal processing and its applications, Penang, Malaysia, pp 464–468

Sáenz-Lechón N, Fraile R, Godino-Llorente JI, Fernández-Baíllo R, Osma-Ruiz V, Gutiérrez-Arriola JM, Arias-Londoño JD (2011) Towards objective evaluation of perceived roughness and breathiness: an approach based on mel-frequency cepstral analysis. Logoped Phoniatr Vocol 36:52–59

Samanta B (2004) Gear fault detection using artificial neural network and vector machines with genetic algorithms. Mech Syst Signal Process 18:625–644

Samanta B, Al-Bushi KR (2003) Artificial neural network based fault diagnostic of rolling elements bearing using time-domain features. Mech Syst Signal Process 17:317–328

Sawalhi N, Randall RB, Endo H (2007) The enhancement of fault detection and diagnosis in rolling element bearings using minimum entropy deconvolution combined with spectral kurtosis. Mech Syst Signal Process 21:2616–2633
Schölkopf B, Smola AJ (2003) A short introduction to learning with kernels. In: Mendelson S, Smola AJ (ed) Proceedings of the machine learning summer school. Springer, Berlin
Shen R, Fu Y, Lu H (2005) A novel image watermarking scheme based on support vector regression. J Syst Software 78:1–8
Tao X, Tao W (2010) Cutting tool wear identification based on wavelet package and SVM. In: Proceedings of the world congress on intelligent control and automation, Shandong, China, pp 5953–5957
Tao XM, Du BX, Xu Y, Wu ZJ (2008) Fault detection for one class of bearings based on AR with self-correlation kurtosis. J Vibr Shock 27:120–124+136
Tellaeche A, Pajares G, Burgos-Artizzu XP, Ribeiro A (2009) A computer vision approach for weeds identification through support vector machines. Appl Soft Comput J 11:908–915
Thissen U, Pepers M, Üstün B, Melssen WJ, Buydens LMC (2004) Comparing support vector machines to PLS for spectral regression applications. Chemometr Intell Lab Syst 73:169–179
Üstün B, Melssen WJ, Buydens LMC (2007) Visualisation and interpretation of support vector regression models. Anal Chim Acta 595:299–309
Vapnik V (1995) The nature of statistical learning theory. Springer, New York
Vapnik V, Lerner A (1963) Pattern recognition using generalized portrait method. Automat Remote Control 24:774–780
Vilakazi CB, Marwala T (2006) Bushing fault detection and diagnosis using extension neural network. In: Proceedings of the 10th IEEE international conference on intelligent engineering systems, London, UK, pp 170–174
Wang CM, Wu MJ, Chen JH, Yu CY (2009) Extension neural network approach to classification of brain MRI. In: Proceedings of the 5th international conference on intelligent information hiding and multimedia signal processing, Kyoto, Japan, pp 515–517
Wang CH, Zhong ZP, Li R, JQ E (2010) Prediction of jet penetration depth based on least square support vector machine. Powder Technol 203:404–411
Wang J, Wang J, Weng Y (2002) Chip design of MFCC extraction for speech recognition. Integr VLSI J 32:111–131
Wang MH (2001) Partial discharge pattern recognition of current transformers using an ENN. IEEE Trans Power Deliv 20:1984–1990
Wang MH, Hung CP (2003) Extension neural network and its applications. Neural Netw 16:779–784
Wang Z, Willett P, DeAguiar PR, Webster Y (2001) Neural network detection of grinding burn from acoustic emission. Int J Mach Tool Manufact 41:283–309
William JH, Davies A, Drake PR (1992) Condition-based maintenance and machine diagnostics. Chapman & Hall, London
Yang BS, Han T, An JL (2004) ART-KOHONEN neural network for fault diagnosis of rotating machinery. Mech Syst Signal Process 18:645–657
Yang BS, Han T, Hwang WW (2005) Fault diagnosis of rotating machinery based on multi-class support vector machines. J Mech Sci Technol 19:846–859
Yang D, Liu P, Wang DQ, Liu HF (2005) Detection of faults and phase-selection using fractal techniques. Autom Electric Power Syst 29:35–39+88
Yang H, Mathew J, Ma L (2005) Fault diagnosis of rolling element bearings using basis pursuit. Mech Syst Signal Process 19:341–356
Yeh CY, Su WP, Lee SJ (2011) Employing multiple-kernel support vector machines for counterfeit banknote recognition. Appl Soft Comput J 11:1439–1447
Zhang J, Qian X, Zhou Y, Deng A (2010) Condition monitoring method of the equipment based on extension neural network. In: Chinese control and decision conference, Taiyan, China, pp 1735–1740
Zhang X, Jiang X, Huang W (2001) Aircraft fault detection based on fractal. J Vibr Shock 20:76–78

Zhao C, Guo Y (2005) Mesh fractal dimension detection on single-phase-to-earth fault in the non-solidly earthed network. In: IEEE power engineering society general meeting, San Francisco, California, pp 752–756

Zhou YP, Jiang JH, Lin WQ, Zou HY, Wu HL, Shen GL, Yu RQ (2006) Boosting support vector regression in QSAR studies of bioactivities of chemical compounds. Eur J Pharm Sci 28:344–353

Chapter 11
On-line Condition Monitoring Using Ensemble Learning

11.1 Introduction

In Chap. 3 the Multi-Layer Perceptron (MLP) neural network was introduced for condition monitoring of a population of cylindrical shells. The MLP technique was explained in detail and after a literature review was conducted the technique was implemented to identify faults in a population of cylindrical shells. In that chapter, modal properties and pseudo-modal energies data were applied to classify faults. The principal component analysis method was applied to reduce the dimensions of the input data. The multifold cross-validation method was used to select the optimal number of hidden units amongst the 20 trained pseudo-modal-energy-networks and the 20 trained modal-property-networks. The pseudo-modal-energy-network and the modal-property-network were found to give similar accuracy in classifying faults.

In Chap. 4, two Bayesian multi-layer perceptron neural networks were developed by applying the hybrid Monte Carlo technique, with one trained using pseudo-modal energies while the other was trained using modal properties. They were then applied to the condition monitoring of a population of cylindrical shells. The pseudo-modal-energy-network gave better results than the modal-property-network.

In Chap. 5, a committee of neural networks technique was presented. It applied pseudo modal energies, modal properties and wavelet transform data simultaneously to identify faults in cylindrical shells. The technique was tested to identify faults in a population of ten steel seam-welded cylindrical shells and could identify faults better than the three individual methods.

Next, Chap. 6 extracted bearing vibration signals features using time-domain fractal-based feature extraction. This method applied the Multi-Scale Fractal Dimension (MFD) which was approximated using the Box-Counting Dimension. The extracted features were then used to classify faults using the Gaussian Mixture Models (GMM) and the hidden Markov Models (HMM). The results showed that the feature extraction method revealed fault specific information. Additionally, the experiment demonstrated that HMM outperformed GMM. Nonetheless, the

T. Marwala, *Condition Monitoring Using Computational Intelligence Methods: Applications in Mechanical and Electrical Systems*, DOI 10.1007/978-1-4471-2380-4_11,

disadvantage of HMM was that it was more computationally expensive to train when compared with the GMM. Consequently, it was concluded that the framework presented gives an improvement in the performance of the bearing fault detection and diagnosis, but it was recommended that the GMM classifier be used when the computational effort is a major issue of consideration.

Chapter 7 presented the application of Fuzzy Set Theory (FST) and fuzzy ARTMAP to diagnose the condition of high voltage bushings. The diagnosis used Dissolved Gas Analysis (DGA) data from bushings based on IEC60599, IEEE C57-104, and the California State University Sacramento (CSUS) criteria for oil impregnated paper (OIP) bushings. FST and fuzzy ARTMAP were compared with regards to accuracy. Both FST and fuzzy ARTMAP could diagnose the bushings condition with 98% and 97.5% accuracy respectively.

Chapter 8 applied the rough set method and the ant colony optimization technique for the condition monitoring of transformer bushings. The theories of rough set and ant colony optimization technique were described and the presented system was tested for the condition monitoring of transformer bushings. The rough set method that was optimized using the ant colony optimization method gave 96.1% accuracy, using 45 rules while the equal-frequency-bin partition model gave 96.4% accuracy, using 206 rules.

In Chap. 9 a technique for fault classification in mechanical systems in the presence of missing data entries was introduced. The technique was based on auto-associative neural networks where the network was trained to recall the input data through some non-linear neural network mapping. From the trained network an error equation with missing inputs as design variables was created. A genetic algorithm was applied to solve for the missing input values. The presented technique was tested on a fault classification problem for a population of cylindrical shells. It was observed that the technique could estimate single-missing-entries to an accuracy of 93% and two-missing-entries to an accuracy of 91%. The approximated values were then applied to the classification of faults and a fault classification accuracy of 94% was observed for single-missing-entry cases and 91% for two-missing-entry cases, while the full database set was able to give a classification accuracy of 96%.

In Chap. 10 feature extraction and condition classification were considered. The feature extraction methods were fractals, Kurtosis and Mel-frequency Cepstral Coefficients. The classification approaches that were applied were the support vector machines (SVM) and extension neural networks (ENN). When applied these techniques gave good results.

Pan et al. (2008) created a remote online machine condition monitoring system which was created using Borland C++ and communication via the internet. A number of signal-processing approaches, for instance time-frequency analysis and order-tracking for signal analysis and pattern recognition were applied using the Borland C++ Builder graphical user interface. The machine fault-diagnostic ability was improved by using the socket application program interface as the transmission control protocol / Internet protocol. The effectiveness of their remote diagnostic system was tested by monitoring a transmission-element test rig and good results were obtained.

Bouhouche et al. (2010) presented a technique for process condition monitoring and evaluation which hybridized the online support vector machine regression and the fuzzy sets approaches. Their technique was based on moving windows so that the past and new data for the model to adapt to the time dependency. A fuzzy analysis was then applied for condition monitoring. Their technique was then applied online to evaluate the quality of a rolling process. The results showed that their technique was simple and gave good results.

Oberholster and Heyns (2009) presented an online condition monitoring technique and applied this to axial-flow turbo-machinery blades. They applied the Eulerian application of laser Doppler vibrometry to accomplish this task. When the method was tested it was found to be viable for the online blade condition monitoring when phase angles at reference frequencies were monitored using a non-harmonic Fourier analysis.

Loutas et al. (2009) presented a condition monitoring system for a single-stage gearbox with induced gear cracks using on-line vibration and acoustic emission measurements. Special attention was paid to the signal processing of the measured vibration and acoustic emission data with the intention of extracting conventional and novel features of diagnostic value from the monitored waveforms. Wavelet-based features used the discrete wavelet transform. The evolution of the chosen features against test time was presented, assessed and the parameters with the most diagnostic characteristics were selected. The advantages of acoustic emission over vibration data for the early diagnosis of natural wear in gear systems were presented.

11.2 Ensemble Methods

The online learning technique implemented in this chapter is based on ensemble learning (Hansen and Salamon 1990; Jordan and Jacobs 1994; Kuncheva et al. 2001). *Ensemble learning* is a technique where multiple models, such as classifiers, are intentionally created and combined to solve a particular problem (Rogova 1994; Polikar 2006). Ensemble learning is usually applied to increase the performance of a model (Xu et al. 1992; Huang and Suen 1993; Dietterich 2000). In this section three ensemble learning approaches are described: bagging, stacking, and adaptive boosting. In particular, the AdaBoost method is described because it was the basis for the creation of the Learn ++ technique, which is the online method adopted for this chapter.

11.2.1 Bagging

Bagging is a technique which is based on the combination of models fitted to randomly selected samples of a training data set to decrease the variance of the prediction model (Efron 1979; Breiman 1996). Bagging basically requires randomly selecting a subset of the training data and using this subset to train a model and repeating this process. Afterwards, all trained models are combined with equal weights to form an ensemble.

11.2.2 Stacking

In the area of modelling, one can choose from a set of models by comparing them using the data that was not used to create the models (Polikar 2006). This prior belief can also be applied to choose a model amongst a set of models, based on a single data set by using a technique called *cross-validation* (Bishop 1995). This is conducted by dividing the data into a *training* data set, which is used to train the models, and a *test* data set. Stacking takes advantage of this prior belief by using the performance from the test data to combine the models instead of choosing among them the best performing model when tested on the test data set (Wolpert 1992).

11.2.3 AdaBoost

Boosting is a method that incrementally creates an ensemble by training each new model with data that the previously trained model misclassified. Then the ensemble, which is a combination of all trained models, is used for prediction.

Adaptive Boosting (AdaBoost) is an extension of boosting to multi-class problems (Freund and Schapire 1997; Schapire et al. 1998). There are many types of AdaBoost, for instance AdaBoost.M1, where each classifier can receive a weighted error of no more than ½, AdaBoost.M2 for those weak classifiers that cannot achieve a weighted error of less than ½.

For AdaBoost.M1, samples are drawn from a distribution D that is updated in such a way that successive classifiers concentrate on difficult cases. This is achieved by adjusting D in such a way that that the earlier, misclassified cases are likely to be present in the following sample. The classifiers are then combined through weighted majority voting. The distribution begins as a uniform distribution so that all cases have equal probability to be drawn into the first data subset S_1.

As described by Polikar (2006), at each iteration t, a new training data subset is sampled, and a weak classifier is trained to create a hypothesis h_t. The error given by this hypothesis with regards to the current distribution is estimated as the sum of distribution weights of the cases misclassified by h_t. AdaBoost.M1 requires that this error is less than ½, and if this requirement is violated then the procedure terminates. The normalized error β_t is then calculated so that the error that is in the [0 0.5] interval is normalized into the [0 1] interval. The transformed error is implemented in the distribution update rule, where $D_t(i)$ is decreased by a factor of $\beta_t, 0 < \beta_t < 1$, if x_i is correctly classified by h_t, or else it is left unaltered. When the distribution is normalized so that $D_{t+1}(i)$ is a proper distribution, the weights of those instances that are misclassified are increased. This update rule guarantees that the weights of all instances correctly classified and the weights of all misclassified instances add up to ½. The requirement for the training error of the base classifier to be less than ½ forces the procedure to correct the error committed by the previous base model. When the training process is complete, the test data are classified by this ensemble of T classifiers, by applying a weighted majority voting procedure where each classifier obtains a voting weight that is inversely proportional to its normalized error (Polikar 2006). The weighted majority voting then selects the class ω allocated the majority vote of all classifiers. The procedure for Adaboost is shown in Algorithm 11.1 (Polikar 2006).

As described by Polikar (2006), the theoretical analysis of the AdaBoost technique shows that the ensemble training error E is bounded above by:

$$E < 2^T \prod_{t=1}^{T} \sqrt{\varepsilon_t \left(1 - \varepsilon_t\right)} \tag{11.1}$$

The $\varepsilon_t < 1/2$ ensemble error E is reduced when new classifiers are added. The AdaBoost is resistant to overfitting, a characteristic that is explained by the margin theory (Schapire 1990; Polikar 2006).

11.3 The Learn++ On-line Method

On-line learning is appropriate for modelling dynamically time-varying systems where the operating conditions change with time. It is also appropriate when the data set available is insufficient and does not completely characterize the system. Another benefit of on-line learning is that it can incorporate new conditions that may be presented by the incoming data.

An on-line bushing condition monitoring system must have incremental learning capability if it is to be used for automatic and continuous on-line monitoring. An on-line bushing monitoring system improves the reliability, diminishes the maintenance cost and minimizes the out-of-service time for a transformer. The basis of on-line learning is incremental learning, which has been studied by a many researchers (Higgins and Goodman 1991; Fu et al. 1996; Yamaguchi et al. 1999; Carpenter et al. 1992). The difficulty in on-line learning is the propensity of an on-line learner to forget the information learned during the initial stages of the learning process (McCloskey and Cohen 1989). The on-line learning technique adopted for this chapter was Learn++ (Polikar et al. 2001).

Vilakazi and Marwala (2007a) applied the on-line incremental learning technique for monitoring the condition of high voltage bushings. Two incremental learning techniques were applied to the problem of condition monitoring. The first technique used was the incremental learning capability of the Fuzzy ARTMAP (FAM), and they investigated whether the ensemble approach can improve the performance

Algorithm 11.1 The AdaBoost algorithm.M1

Input:

- Training data $X = \{\mathbf{x}_1, \mathbf{x}_2, ..., \mathbf{x}_n\}$ with correct labels $\Delta = \{y_1, y_2, ..., y_n\}$
- Weak learn algorithm, known as **Weaklearn**
- Integer *T*, speciying the number of classifiers

$$D_1(i) = 1/n; i = 1, ..., n$$

For $t = 1, 2, \ldots, T$;

1. Sample a training subset S_t, according to the distribution D_t
2. Train **Weaklearn** with S_t, receive hypothesis $h_t : X \rightarrow \Delta$
3. Estimate the error of h_t : $\varepsilon_t = \sum_{i=1}^{n} I\,[h_t(\mathbf{x}_i) \neq y_i] \cdot D_t(i) = \sum_{t:h_t(\mathbf{x}_i) \neq y_i} D_t(i)$ If $\varepsilon_t > \frac{1}{2}$ terminate.
4. Estimate the normalized error $\beta_t = \varepsilon_t / (1 - \varepsilon_t) \Rightarrow 0 \leq \beta_t \leq 1$
5. Update the distribution D_t: $D_{t+1}(i) = \frac{D_t(i)}{Z_t} \times \begin{cases} \beta_t, & \textit{if } h_t(\mathbf{x}_i) y_i \\ 1, & \textit{otherwise} \end{cases}$ where Z_t is the normalization constant so that D_{t+1} becomes a proper distribution function.

Test using majority voting given an unlabeled example *z* as follows:

1. Count the total vote from the classifiers $V_j = \sum_{t:h_t(z)} \log(1/\beta_t) j = 1, ..., C$
2. Select the class that receives the highest number of votes as the final classification.

of the FAM. The second technique applied was Learn++ that implemented an ensemble of the multi-layer perceptron classifiers. Both methods were performed well when tested for transformer bushing condition monitoring.

Mohamed et al. (2007) applied incremental learning for the classification of protein sequences. They used the fuzzy ARTMAP as an alternative machine learning system with the ability to incrementally learn new data as it becomes available. The fuzzy ARTMAP was seen to be comparable to many other machine learning systems. The application of an evolutionary strategy in the selection and combination of individual classifiers into an ensemble system, coupled with the incremental learning capability of the fuzzy ARTMAP was shown to be suitable as a pattern classifier. Their algorithm was tested using the data from the G-Coupled Protein Receptors Database and it demonstrated a good accuracy of 83%.

Mohamed et al. (2006) applied fuzzy ARTMAP for multi-class protein sequence classification. They presented a classification system that used pattern recognition method to produce a numerical vector representation of a protein sequence and then classified the sequence into a number of given families. They applied fuzzy ARTMAP classifiers and showed that, when coupled with a genetic algorithm based feature subset selection, the system could classify protein sequences with an accuracy of 93%. This accuracy was then compared to other classification techniques and it was shown that the fuzzy ARTMAP was most suitable because of its high accuracy, quick training times and ability to incrementally learn.

Perez et al. (2010) applied a population-based, incremental learning approach to microarray gene expression feature selection. They evaluated the usefulness of the Population-Based Incremental Learning (PBIL) procedure in identifying a class differentiating gene set for sample classification. PBIL was based on iteratively evolving the genome of a search population by updating a probability vector, guided by the extent of class-separability demonstrated by a combination of features. The PBIL was then compared to standard Genetic Algorithm (GA) and an Analysis of Variance (ANOVA) method. The procedures were tested on a publically available three-class leukaemia microarray data set ($n = 72$). After running 30 repeats of both GA and PBIL, the PBIL could identify an average feature-space separability of 97.04%, while GA achieved an average class-separability of 96.39%. The PBIL also found smaller feature-spaces than GA, (PBIL – 326 genes and GA – 2652) thus excluding a large percentage of redundant features. It also, on average, outperformed the ANOVA approach for $n = 2652$ (91.62%), $q < 0.05$ (94.44%), $q < 0.01$ (93.06%) and $q < 0.005$ (95.83%). The best PBIL run (98.61%) even outperformed ANOVA for $n = 326$ and $q < 0.001$ (both 97.22%). PBIL's performance was credited to its ability to direct the search, not only towards the optimal solution, but also away from the worst.

Hulley and Marwala (2007) applied GA-based incremental learning for optimal weight and classifier selection. They then compared Learn++, which is an incremental learning algorithm to the new Incremental Learning Using Genetic Algorithm (ILUGA). Learn++ demonstrated good incremental learning capabilities on benchmark datasets on which the new ILUGA technique was tested. ILUGA showed good incremental learning ability using only a few classifiers and did not suffer from catastrophic forgetting. The results obtained for ILUGA on the Optical Character Recognition (OCR) and Wine datasets were good, with an overall accuracy of 93% and 94% respectively showing a 4% improvement over Learn++.MT for the difficult multi-class OCR dataset.

Lunga and Marwala (2006a) applied a time series analysis using fractal theory and on-line ensemble classifiers to model the stock market. The fractal analysis was implemented as a concept to identify the degree of persistence and self-similarity within the stock market data. This concept was carried out using the Rescaled range analysis (R/S) technique. The R/S analysis outcome was then applied to an on-line incremental algorithm (Learn++) that was built to classify the direction of movement of the stock market. The use of fractal geometry in this study provided a way of determining quantitatively the extent to which the time series data could be predicted. In an extensive test, it was demonstrated that the R/S analysis provided a very sensitive technique to reveal hidden long runs and short run memory trends within the sample data. A time series data that was measured to be persistent was used in training the neural network. The results from the Learn++ algorithm showed a very high level of confidence for the neural network to classify sample data accurately.

Lunga and Marwala (2006b) applied incremental learning for the on-line forecasting of stock market movement direction. In particular, they presented a specific application of the Learn++ algorithm, and investigated the predictability of financial movement direction with Learn++ by forecasting the daily movement direction of the Dow Jones. The framework was implemented using the Multi-Layer Perceptron (MLP) as a weak learner. First, a weak learning algorithm, which tried to learn a class concept with a single input perceptron, was established. The Learn++ algorithm was then applied to improve the weak MLP learning capacity and thus introduced the concept of incremental on-line learning. The presented framework could adapt as new data were introduced and could classify the data well.

Vilakazi and Marwala (2007b) applied incremental learning to bushing condition monitoring. They presented a technique for bushing fault condition monitoring using the fuzzy ARTMAP. The fuzzy ARTMAP was introduced for bushing condition monitoring because it can incrementally learn information as it becomes available. An ensemble of classifiers was used to improve the classification accuracy of the systems. The test results showed that the fuzzy ARTMAP ensemble gave an accuracy of 98.5%. In addition, the results showed that the fuzzy ARTMAP could update its knowledge in an incremental fashion without forgetting the previously learned information.

Nelwamondo and Marwala (2007) successfully applied a technique for handling missing data from heteroskedastic and non-stationary data. They presented a computational intelligence approach for predicting missing data in the presence of concept drift using an ensemble of multi-layered feed-forward neural networks. Six instances prior to the occurrence of missing data were used to approximate the missing values. The algorithm was applied to a simulated time series data sets that resembled non-stationary data from a sensor. Results showed that the prediction of missing data in a non-stationary time series data was possible but was still a challenge. For one test, up to 78% of the data could be predicted within a 10% tolerance range of accuracy.

Other successful implementations of incremental learning techniques include its use in anomaly detection (Khreich et al. 2009), in human robot interaction (Okada et al. 2009), for online handwriting recognition (Almaksour and Anquetil 2009), for predicting human and vehicle motion (Vasquez et al. 2009) and in visual learning (Huang et al. 2009).

11.3.1 Learn++

Learn++ is an incremental learning algorithm that was introduced by Polikar and co-workers (Polikar et al. 2000, 2001, 2002; Muhlbaier et al. 2004; Erdem et al. 2005; Polikar 2006). It is based on AdaBoost and applies multiple classifiers to enable the system to learn incrementally. The algorithm operates on the concept of using many classifiers that are weak learners to give a good overall classification. The weak learners are trained on a separate subset of the training data and then the classifiers are combined using a weighted majority vote. The weights for the weighted majority vote are chosen using the performance of the classifiers on the entire training dataset.

Each classifier is trained using a training subset that is drawn according to a specified distribution. The classifiers are trained using a weak learn algorithm (WeakLearn). The requirement for the WeakLearn algorithm is that it must give a classification rate of less than 50% initially (Polikar et al. 2002). For each database *Dk* that contains training sequence, *S*, where *S* contains learning examples and their corresponding classes, Learn++ starts by initializing the weights, *w*, according to a specified distribution *DT*, where *T* is the number of hypothesis. Firstly the weights are initialized to be uniform, thereby giving equal probability for all cases selected for the first training subset and the distribution is given by (Polikar et al. 2002):

$$D = {}^{1}\!/_{m} \tag{11.2}$$

Here, *m* represents the number of training examples in *S*. The training data are then divided into training subset *TR* and testing subset *TE* to ensure the WeakLearn capability. The distribution is then used to select the training subset *TR* and testing subset *TE* from *Sk*. After the training and testing subset have been selected, the WeakLearn algorithm is implemented. The WeakLearner is trained using subset *TR*. A hypothesis, h_t, obtained from a WeakLearner is tested using both the training and testing subsets to obtain an error (Polikar et al. 2002):

$$\varepsilon_t = \sum_{t:h_i(x_i) \neq y_i} D_t(i) \tag{11.3}$$

The error is required to be less than 0.5; a normalized error βt is computed using (Polikar et al. 2002):

$$B_t = {}^{\varepsilon_t}\!/_{1 - \varepsilon_t} \tag{11.4}$$

If the error is greater than 0.5, the hypothesis is discarded and a new training and testing subsets are selected according to a distribution D_T and another hypothesis is computed. All classifiers generated are then combined using weighted majority voting to obtain the composite hypothesis, H_t (Polikar et al. 2002):

$$H_t = \arg\max_{y \in Y} \sum_{t:h_t(x)=y} \log\left({}^{1}\!/_{\beta_t}\right) \tag{11.5}$$

Weighted majority voting gives higher voting weights to a hypothesis that performs well on the training and testing data subsets. The error of the composite hypothesis is computed as follows (Polikar et al. 2002):

$$E_t = \sum_{t:H_i(x_i)\neq y_i} D_t(i) \tag{11.6}$$

If the error is greater than 0.5, the current hypothesis is discarded and the new training and testing data are selected according to the distribution D_T. Otherwise, if the error is less than 0.5, then the normalized error of the composite hypothesis is computed as follows (Polikar et al. 2002):

$$B_t = {}^{E_t}/_{1 - E_t} \tag{11.7}$$

The error is used in the distribution update rule, where the weights of the correctly classified case are reduced, consequently increasing the weights of the misclassified instances. This ensures that the cases that were misclassified by the current hypothesis have a higher probability of being selected for the subsequent training set. The distribution update rule is given by the following equation (Polikar et al. 2002):

$$w_{t+1} = w_t(i) \times B_t^{1-[|H_t(x_i)\neq y_i|]} \tag{11.8}$$

After the T hypothesis has been created for each database, the final hypothesis is computed by combining the hypotheses using weighted majority voting as described by the following equation (Polikar et al. 2002):

$$H_t = \arg\max_{y\in Y} \sum_{k=1}^{K} \sum_{t:H_t(x)=y} \log\left({}^{1}/_{\beta_t}\right) \tag{11.9}$$

The Learn++ algorithm is represented diagrammatically in Fig. 11.1.

11.3.2 Confidence Measurement

A technique is used to estimate the confidence of the algorithm about its own decision. A majority of hypotheses agreeing on given instances can be interpreted as an indication of confidence on the decision proposed. If it is assumed that a total of T hypotheses are generated in k training sessions for a C-class problem, then for any given example, the final classification class, the total vote class c receives is given by (Muhlbaier et al. 2004):

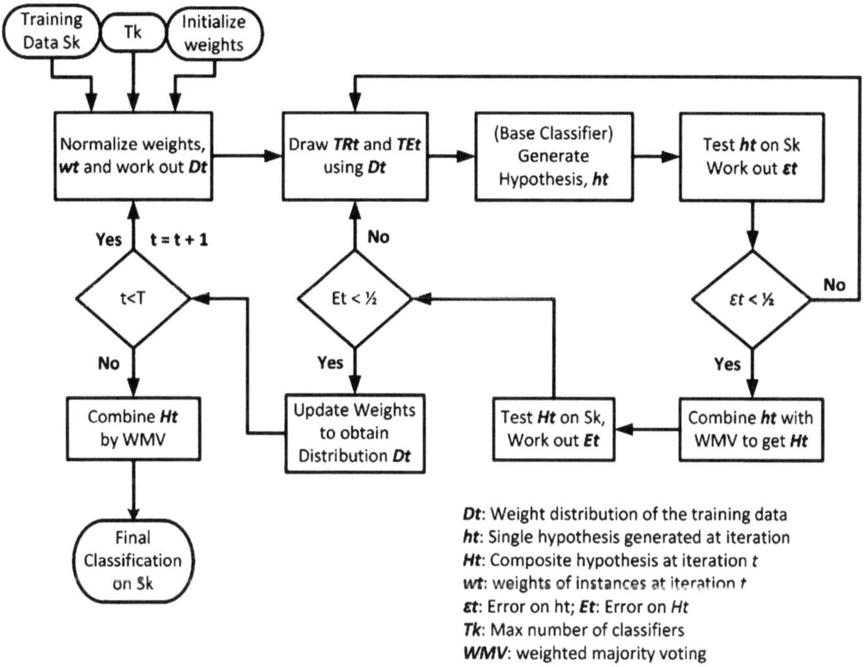

Fig. 11.1 Block diagram of a Learn++ algorithm

$$\zeta_c = \sum_{t:h_t(\mathbf{x})=c} \Psi_t \tag{11.10}$$

where Ψt denotes the voting weights of the tth, hypothesis h_t.

Normalizing the votes received by each class can be performed as follows (Muhlbaier et al. 2004):

$$\lambda_c = \frac{\zeta_c}{\sum_{c=1}^{C} \zeta_c} \tag{11.11}$$

Here, λc can be interpreted as a measure of confidence on a scale of 0–1. A high value of λc shows high confidence in the decision and conversely, a low value of λc shows low confidence in the decision. It should be noted that the λc value does not represent the accuracy of the results, but the confidence of the system in its own decision.

11.4 Multi-Layer Perceptrons

The architecture considered in this chapter to create the WeakLearn was the multi-layer perceptron (MLP) as described in great detail in Chap. 3. The MLP can be defined as a feed-forward neural network model that approximates the relationship between a set of input data and a set of appropriate output data. Its foundation is the standard linear perceptron. It makes use of three or more layers of neurons usually with non-linear activation functions. This is because it can distinguish data that are not linearly separable, or separable by a hyper-plane. The multi-layer perceptron has been used to model many complex systems in areas such as mechanical and aerospace engineering as well as for modelling interstate conflict (Marwala 2007; Marwala 2009; Marwala 2010; Marwala and Lagazio 2011).

The MLP neural network consists of multiple layers of computational units, usually inter-connected in a feed-forward way (Bishop 1995). Each neuron in one layer is directly connected to the neurons of the subsequent layer. A fully connected two-layered MLP architecture was used for this chapter. A two-layered MLP architecture was used because of the universal approximation theorem, which states that a two-layered architecture is adequate for MLP and, consequently, it can approximate data of arbitrary complexity (Bishop 1995).

11.5 Experimental Investigation

A dissolved gas analysis is used to estimate the faulty gases in bushing oil. The information from the dissolved gas analysis reflects the states of the transformer and bushing. Ten diagnostic gases are extracted: CH_4, C_2H_6, C_2H_4, C_2H_2, H_2, CO, CO_2, N_2, O_2 and total dissolved combustible gases. The total dissolved combustible gas is given by the sum of methane, hydrogen, acetylene, ethane, ethylene and hydrogen. The faulty gases are analyzed using the IEEE C57.104 standards. Data pre-processing is an integral part of neural network architecture. Data pre-processing makes it easier for the network to learn. Data are normalized to fall within 0 and 1.

The first experiment evaluated the incremental capability of the Learn++ algorithm using a first-level fault diagnosis, which was aimed at classifying the presence or the absence of faults in transformer bushings. The data used were collected from bushings over a period of 2.5 years from bushings in service. The algorithm was implemented with 1,500 training examples and 4,000 validation examples. The training data were divided into five databases each with 300 training instances. In each training session, Learn++ was provided with each database and 20 hypotheses were generated. The WeakLearner used an MLP with 10 input layer neurons, 5 hidden layer neurons and one output layer neuron. To ensure that the technique retained previously learned data, the previous database was tested at each training session.

Table 11.1 Performance of Learn++ for first level online condition monitoring (Key: S = dataset)

Dataset	S_1	S_2	S_3	S_4	S_5
S_1	89.5	85.8	83.0	86.9	85.3
S_2	–	91.4	94.2	93.7	92.9
S_3	–	–	93.2	90.1	91.4
S_4	–	–	–	92.2	94.5
S_5	–	–	–	–	98.0
Learn ++ Testing (%)	65.7	79.0	85.0	93.5	95.8

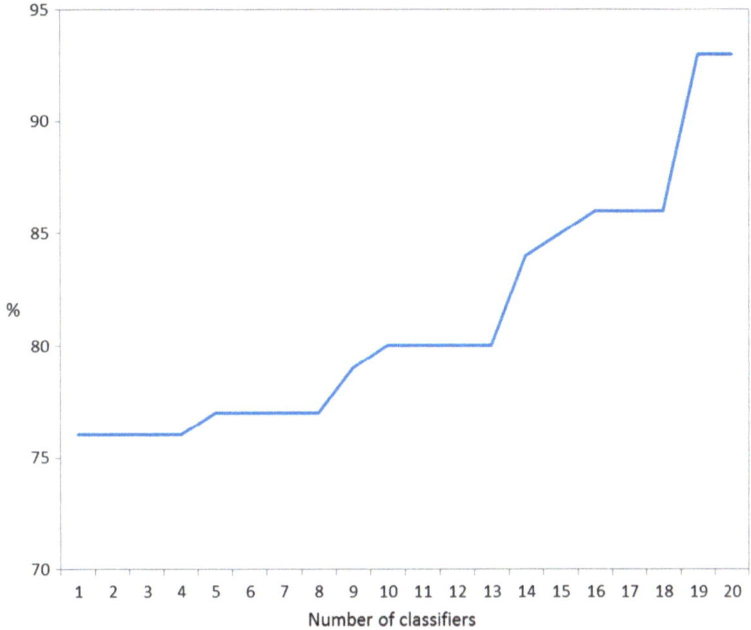

Fig. 11.2 Performance of Learn++ on training data against the number of classifiers

The first row of Table 11.1 shows the performance of Learn++ on the training data for different databases. On average, the WeakLearner gave 60% classification rate on its training dataset, which improved to 98% when the hypotheses were combined.

These results show the performance improvement of Learn++ with a single database. Each column shows the performance of current and previous databases. This is to indicate that Learn++ did not forget the previously learned information when new data were introduced.

The classifiers' performance on the testing dataset steadily increased from 65.7% to 95.8% as new databases became available, demonstrating the incremental capability of Learn++ as shown in Fig. 11.2.

A second experiment was performed to evaluate whether the frameworks can accommodate new classes. The results appear in Table 11.2. The faulty data

Table 11.2 Performance of Learn++ for second stage bushing condition monitoring

Dataset	S_1	S_2	S_3	S_4	S_5
S_1	95.0	95.2	94.6	95.7	95.1
S_2	–	96.3	96.0	96.8	95.3
S_3	–	–	97.0	96.4	96.5
S_4	–	–	–	97.8	96.8
S_5	–	–	–	–	99.2
Learn ++ Testing (%)	60.0	65.2	76.0	83.0	95.3

were divided into 1,000 training examples and 2,000 validation examples, which contained all three classes. The training data were divided into five databases, each with 200 training instances. The first and second databases contained training examples of partial discharges and thermal faults.

The data with unknown faults were introduced in training session three. In each training session, Learn++ was provided with each database and 20 hypotheses were generated. The classifiers performance increased from 60% to 95.3% as new classes were introduced in subsequent training datasets. The final experiment addressed the problem of bushing condition monitoring using a MLP network that was trained using batch learning. This was done to compare the classification rate of Learn++ with that of a MLP.

A MLP with the same set of training example as Learn++ was trained and the trained MLP was tested with the same validation data as Learn++. This test was conducted for the first and second levels of fault classification. In the first level fault diagnosis, the MLP gave a classification rate of 97.2% whereas the second level MLP gave a classification rate of 96.0%. This was when the classifier had seen all the fault classes *a priori*. If the classifier had not seen all the fault cases, the performance decreased from 65.7% for database 1–30.0% for databases 2–3 for the first level fault classification.

11.6 Conclusion

This chapter presented an on-line bushing condition monitoring approach, which can adapt to newly acquired data. This technique was capable of factoring into account new classes that were introduced by incoming data and was implemented using an incremental learning algorithm that used the MLP called Learn++. The test results improved from 67.5% to 95.8% as new data were introduced and improved from 60% to 95.3% as new conditions were introduced. On average, the confidence value of the framework about its decision was 0.92.

References

Almaksour A, Anquetil E (2009) Fast incremental learning strategy driven by confusion reject for online handwriting recognition. In: Proceedings of the international conference on document analysis and recognition, Barcelona, Spain, pp 81–85

Bishop CM (1995) Neural networks for pattern recognition. Oxford University Press, Oxford

Bouhouche S, Yazid LL, Hocine S, Bast J (2010) Evaluation using online support-vector-machines and fuzzy reasoning. Application to condition monitoring of speeds rolling process. Control Eng Pract 18:1060–1068

Breiman L (1996) Bagging predictors. Mach Learn 24:123–140

Carpenter GA, Grossberg S, Marhuzon N, Reynolds JH, Rosen DB (1992) ARTMAP: a neural network architecture for incremental learning supervised learning of analog multidimensional maps. IEEE Trans Neural Netw 3:698–713

Dietterich TG (2000) Ensemble methods in machine learning. Lect Notes Comput Sci 1857:1–15

Efron B (1979) Bootstrap methods: another look at the jackknife. Ann Stat 7:1–26

Erdem Z, Polikar R, Gurgen F, Yumusak N (2005) Reducing the effect of out-voting problem in ensemble based incremental support vector machines. Lect Notes Comput Sci 3697:607–612

Freund Y, Schapire RE (1997) Decision-theoretic generalization of on-line learning and an application to boosting. J Comput Syst Sci 55:119–139

Fu L, Hsu HH, Principe JC (1996) Incremental backpropagation networks. IEEE Trans Neural Netw 7:757–761

Hansen LK, Salamon P (1990) Neural network ensembles. IEEE Trans Pattern Anal Mach Intell 12:993–1001

Higgins CH, Goodman RM (1991) Incremental learning for rule based neural network. In: Proceedings of the internatinal joint conference on neural networks, Seattle, Washington, pp 875–880

Huang YS, Suen CY (1993) Behavior-knowledge space method for combination of multiple classifiers. In: Proceedings of the IEEE computer vision and pattern recognition conference, New York, USA, pp 347–352

Huang D, Yi Z, Pu X (2009) A new incremental PCA algorithm with application to visual learning and recognition. Neural Process Lett 30:171–185

Hulley G, Marwala T (2007) Genetic algorithm based incremental learning for optimal weight and classifier selection. In: Proceedings of the AIP conference, Vancouver, Canada, pp 258–267

Jordan MJ, Jacobs RA (1994) Hierarchical mixtures of experts and the EM algorithm. Neural Comput 6:181–214

Khreich W, Granger E, Miri A, Sabourin RA (2009) A comparison of techniques for on-line incremental learning of HMM parameters in anomaly detection. In: Proceedings of the IEEE symposium on computational intelligence for security and defense applications, Ottawa, pp 1–8

Kuncheva LI, Bezdek JC, Duin R (2001) Decision templates for multiple classifier fusion: an experimental comparison. Pattern Recognit 34:299–314

Loutas TH, Sotiriades G, Kalaitzoglou I, Kostopoulos V (2009) Condition monitoring of a single-stage gearbox with artificially induced gear cracks utilizing on-line vibration and acoustic emission measurements. Appl Acoust 70:1148–1159

Lunga D, Marwala T (2006a) Online forecasting of stock market movement direction using the improved incremental algorithm. Lect Notes Comput Sci 4234:440–449

Lunga D, Marwala T (2006b) Time series analysis using fractal theory and online ensemble classifiers. Lect Notes Comput Sci 4304:312–321

Marwala T (2007) Computational intelligence for modelling complex systems. Research India Publications, Delhi

Marwala T (2009) Computational intelligence for missing data imputation, estimation and management: knowledge optimization techniques. IGI Global Publications, New York

Marwala T (2010) Finite element model updating using computational intelligence techniques. Springer, London

Marwala T, Lagazio M (2011) Militarized conflict modeling using computational intelligence techniques. Springer, London
McCloskey M, Cohen N (1989) Catastrophic interference connectionist networks: the sequential learning problem. Psychol Learn Motiv 24:109–164
Mohamed S, Rubin D, Marwala T (2006) Multi-class protein sequence classification using fuzzy ARTMAP. In: Proceedings of the IEEE international conference on systems, man, and cybernetics, Taipei City, pp 1676–1681
Mohamed S, Rubin D, Marwala T (2007) Incremental learning for classification of protein sequences. In: Proceedings of the IEEE international joint conference on neural networks, Orlando, Florida, pp 19–24
Muhlbaier M, Topalis A, Polikar R (2004) Learn++.MT: a new approach to incremental learning. Lect Notes Comput Sci 3077:52–61
Nelwamondo FV, Marwala T (2007) Handling missing data from heteroskedastic and nonstationary data. Lect Notes Comput Sci 4491:1293–1302
Oberholster AJ, Heyns PS (2009) Online condition monitoring of axial-flow turbomachinery blades using rotor-axial Eulerian laser Doppler vibrometry. Mech Syst Signal Process 23: 1634–1643
Okada S, Kobayashi Y, Ishibashi S, Nishida T (2009) Incremental learning of gestures for human-robot interaction. AI Soc 25:155–168
Pan MC, Li PC, Cheng YR (2008) Remote online machine condition monitoring system. Measurement 41:912–921
Perez M, Featherston J, Marwala T, Scott LE, Stevens DM (2010) A population-based incremental learning approach to microarray gene expression feature selection. In: Proceedings of the IEEE 26th convention of electrical and electronic engineers, Eilat, Israel, pp 10–14
Polikar R (2006) Ensemble based systems in decision making. IEEE Circuit Syst Mag 6:21–45
Polikar R, Udpa L, Udpa S, Honavar V (2000) Learn++: an incremental learning algorithm for multilayer perceptrons. In: Proceedings of IEEE 25th international conference on acoustics, speech and signal processing, Michigan, USA, pp 3414–3417
Polikar R, Udpa L, Udpaand S, Honavar V (2001) Learn++: an incremental learning algorithm for supervised neural networks. IEEE Trans Syst Man Cybern 31:497–508
Polikar R, Byorick J, Krause S, Marino A, Moreton M (2002) Learn++: a classifier independent incremental learning algorithm for supervised neural networks. In: Proceedings of the international joint conference on neural networks, Honolulu, pp 1742–1747
Rogova G (1994) Combining the results of several neural network classifiers. Neural Netw 7: 777–781
Schapire RE (1990) The strength of weak learnability. Mach Learn 5:197–227
Schapire RE, Freund Y, Bartlett P, Lee WS (1998) Boosting the margin: a new explanation for the effectiveness of voting methods. Ann Stat 26:51–1686
Vasquez D, Fraichard T, Laugier C (2009) Growing hidden Markov models: an incremental tool for learning and predicting human and vehicle motion. Int J Robot Res 28:1486–1506
Vilakazi CB, Marwala T (2007a) Incremental learning and its application to bushing condition monitoring. Lect Notes Comput Sci 4491:1237–1246
Vilakazi CB, Marwala T (2007b) Online incremental learning for high voltage bushing condition monitoring. In: Proceedings of the international joint conference on neural networks, Orlando, Florida, pp 2521–2526
Wolpert DH (1992) Stacked generalization. Neural Netw 5:241–259
Xu L, Krzyzak A, Suen CY (1992) Methods for combining multiple classifiers and their applications to handwriting recognition. IEEE Trans Syst Man Cybern 22:418–435
Yamaguchi K, Yamaguchi N, Ishii N (1999) Incremental learning method with retrieving of interfered patterns. IEEE Trans Neural Netw 10:1351–1365

Chapter 12
Conclusion

12.1 Introduction

In Chap. 1, condition monitoring methods in mechanical and electrical systems were reviewed. A condition monitoring framework was discussed which entailed the domain in which the data were visualized and in particular the time, modal, frequency and time-frequency domains. A generalized condition monitoring framework which encompasses the data acquisition device, data analysis device, feature selection device, and decision making device was presented. Techniques for decision making devices were introduced: finite element models, correlation based methods and computational intelligence techniques.

Chapter 2 reviewed data processing methods for condition monitoring in mechanical and electrical systems. Approaches for acquiring data were described and techniques for analyzing data were explained. In particular, modal properties, pseudo-modal energies, wavelet and Mel-frequency Cepstral Coefficients techniques were described. Furthermore, the principal component analysis technique was described. In conclusion, examples that were followed for this book were described. These examples were gearbox data, a population of cylindrical shells data and transformer bushing data. Additionally, Bayesian neural networks were implemented using the hybrid Monte Carlo method and trained using pseudo-modal energies and modal properties data. The pseudo-modal-energy-network was found to offer better results than the modal-property-network.

In Chap. 3 the multi-layer perceptron neural network was introduced for condition monitoring of a population of cylindrical shells. The multi-layer perceptron method was explained in detail and a literature review of the pertinent literature was presented and applied to identify faults in a population of cylindrical shells. The modal properties and pseudo-modal energies data were used to classify faults and a principal component analysis was applied to reduce the dimensions of the input data. The pseudo-modal-energy-network and the modal-property-network were found to give similar levels of accuracy in classifying faults.

T. Marwala, *Condition Monitoring Using Computational Intelligence Methods: Applications in Mechanical and Electrical Systems*, DOI 10.1007/978-1-4471-2380-4_12,

In Chap. 4, two Bayesian multi-layer perceptron neural networks were developed by applying the hybrid Monte Carlo technique with one trained using the pseudo-modal energies while the other trained using the modal properties. They were then applied for condition monitoring of a population of cylindrical shells. The pseudo-modal-energy-network gave better results than the modal-property-network.

In Chap. 5, a committee of neural networks method applied the pseudo modal energies, modal properties and wavelet transform data simultaneously to identify faults in cylindrical shells. When tested to identify faults it could identify faults better than the three individual methods.

In Chap. 6, bearing vibration signals features were extracted using the time-domain fractal-based feature extraction and the extracted features were then applied to classify faults using Gaussian Mixture Models (GMMs) and hidden Markov Models (HMMs). The HMM was found to out-perform GMM.

In Chap. 7, the Fuzzy Set Theory (FST) and fuzzy ARTMAP was applied to diagnose the condition of high voltage bushings. Both FST and fuzzy ARTMAP could diagnose the bushing's condition.

In Chap. 8, the rough set technique and ant colony optimization methods were applied for the condition monitoring of transformer bushings. The rough set optimized using the ant colony optimization method gave better results than the equal-frequency-bin partition model.

In Chap. 9, a method was introduced for fault classification in mechanical systems in the presence of missing data entries, based on auto-associative neural networks and a genetic algorithm. The approximated values were then applied for the classification of faults. A fault classification accuracy of 94% was observed for single-missing-entry cases and 91% for two-missing-entry cases, while the full database set gave a classification accuracy of 96%.

In Chap. 10, feature extraction methods: fractals, Kurtosis and Mel-frequency Cepstral Coefficients were used as inputs to the Support vector machine (SVM) and extension neural network (ENN) and then applied for condition monitoring. When tested for condition monitoring they were observed to give good results.

In Chap. 11, an on-line bushing condition monitoring technique was introduced, which was capable of adapting to newly acquired data. This method could accommodate new classes that were introduced by incoming data. It was implemented using an incremental learning algorithm that used the multi-layered perceptron. The testing results improved from 67.5% to 95.8% as new data were introduced and the testing results improved from 60% to 95.3% as new conditions were introduced.

12.2 Future Studies

A multi-agent system uses intelligent software that can learn, adapt and act independently to solve complex problems. For example, Marwala and Hurwitz (2009) applied a multi-agent system to design software that can bluff. Elsewhere, Pereira et al. (2001) applied a multi-agent system to simulate a trading system.

Marwala et al. (2001) applied a multi-agent system for stock market prediction while Abramov et al. (2001) studied the ontological basis of a multi-agent system. Teweldemedhin et al. (2004) applied a multi-agent system to study the infection characteristics of the HIV epidemic. Elsewhere, Marivate et al. (2008) applied a multi-agent system for designing a neural network based recommender system. In the future an integrated study on how the multi-agent systems can be used in condition monitoring should be explored.

The ability of a classifier to take on new information and classes by evolving the classifier system without it having to be fully retrained is known as *incremental learning*. In this book an incremental learning system, known as Learn++, was successfully applied to condition monitoring. For future studies a new method called Incremental Learning Using a Genetic Algorithm (ILUGA) which was successfully applied to Optical Character Recognition (OCR) and Wine datasets could be applied to condition monitoring and be compared to the Learn++ system (Hulley and Marwala 2007).

In this book, a committee of learning networks was applied to condition monitoring and was found to perform better than the individual methods. For future studies, a committee method should be adapted with the ability to autonomously evolve (Marwala 2009, 2010).

In this book, rough sets theory was applied for condition monitoring. For future work a neuro-rough model can be a combination of neural networks with rough set theory for condition monitoring (Marwala and Crossingham 2008; Marwala and Lagazio 2011). This model will naturally combine the accuracy of the neural network with the transparency of the rough set model.

References

Abramov VA, Szirbik NB, Goossenaerts JBM, Marwala T, de Wilde P, Correia L, Mariano P, Ribeiro R (2001) Ontological basis for open distributed multi-agent system. In: Proceedings of the symposium on adaptive agents and multi-agent systems, York, UK, pp 33–43

Hulley G, Marwala T (2007) Genetic algorithm based incremental learning for optimal weight and classifier selection. Comput Model Life Sci 952:258–267

Marivate V, Ssali G, Marwala T (2008) An intelligent multi-agent recommender system for human capacity building. In: Proceedings of the 14th IEEE mediterranean electrotechnical conference, Ajaccio, France, pp 909–915

Marwala T (2009) Computational intelligence for missing data imputation, estimation and management: knowledge optimization techniques. IGI Global Publications, New York

Marwala T (2010) Finite element model updating using computational intelligence techniques. Springer, London

Marwala T, Crossingham B (2008) Neuro-rough models for modelling HIV. In: Proceedings of the IEEE international conference on man, systems, and cybernetics, Singapore, pp 3089–3095

Marwala T, Hurwitz E (2009) A multi-agent approach to bluffing. In: Ahmed S, Karsiti MN (eds) Multiagent systems. InTech Open Access Publisher, Vienna

Marwala T, Lagazio M (2011) Militarized conflict modeling using computational intelligence techniques. Springer, London

Marwala T, de Wilde P, Correia L, Mariano P, Ribeiro R, Abramov V, Szirbik N, Goossenaerts J (2001) Scalability and optimisation of a committee of agents using a genetic algorithm. In: Proceedings of the international symposium on soft computing and intelligent systems for industry, Scotland, UK, pp 1–6

Pereira A, Correia L, Ribeiro R, Abramov V, Szirbik N, Goossenaerts J, Marwala T, de Wilde P, Mariano P (2001) Simulation of a trading multi-agent system. In: Proceedings of the IEEE international conference on systems, man, and cybernetics, Tucson, Arizona, pp 3378–3384

Teweldemedhin E, Marwala T, Mueller C (2004) Agent-based modelling: a case study in HIV epidemic. In: Proceedings of the IEEE 4th international conference in hybrid intelligent systems, Kitakyushu, Japan, pp 154–159

Biography

Tshilidzi Marwala was born in Venda (Limpopo, South Africa) and is currently the Dean of Engineering at the University of Johannesburg. Previously he was a Full Professor of Electrical Engineering, the holder of the Carl and Emily Fuchs Chair of Systems and Control Engineering, as well as the holder of the DST/NRF South Africa Research Chair of Systems Engineering at the University of the Witwatersrand. He is a professor extraordinaire at the University of Pretoria and is on the boards of directors of Denel, EOH and City Power Johannesburg. He is a Fellow of the following institutions: the Council for Scientific and Industrial Research, the South African Academy of Engineering, the South African Academy of Science, TWAS – The Academy of Science for the Developing World and the Mapungubwe Institute for Strategic Reflection. He is a Senior Member of the IEEE and a Distinguished Member of the ACM. He is a Trustee of the Bradlow Foundation as well as the Carl and Emily Fuchs Foundation. He is the recipient of the Order of Mapungubwe. He holds a Bachelor of Science in Mechanical Engineering (*Magna Cum Laude*) from Case Western Reserve University, a Master of Engineering from the University of Pretoria, and a Ph.D. in Engineering from the University of Cambridge. He was a post-doctoral research associate at Imperial College (London) and was a visiting fellow at Harvard University and Wolfson College (Cambridge). His research interests include the application of computational intelligence to engineering, computer science, finance, social science and medicine. He has successfully supervised 36 masters and 6 Ph.D. students, published over 220 refereed papers, holds 3 patents and has authored 5 books.

T. Marwala, *Condition Monitoring Using Computational Intelligence Methods: Applications in Mechanical and Electrical Systems*, DOI 10.1007/978-1-4471-2380-4,

Index

T. Marwala, *Condition Monitoring Using Computational Intelligence Methods: Applications in Mechanical and Electrical Systems*, DOI 10.1007/978-1-4471-2380-4,